T0230455

SpringerBriefs in Applied Sciences and Technology

SpringerBriefs present concise summaries of cutting-edge research and practical applications across a wide spectrum of fields. Featuring compact volumes of 50 to 125 pages, the series covers a range of content from professional to academic.

Typical publications can be:

- A timely report of state-of-the art methods
- An introduction to or a manual for the application of mathematical or computer techniques
- A bridge between new research results, as published in journal articles
- A snapshot of a hot or emerging topic
- An in-depth case study
- A presentation of core concepts that students must understand in order to make independent contributions

SpringerBriefs are characterized by fast, global electronic dissemination, standard publishing contracts, standardized manuscript preparation and formatting guidelines, and expedited production schedules.

On the one hand, **SpringerBriefs in Applied Sciences and Technology** are devoted to the publication of fundamentals and applications within the different classical engineering disciplines as well as in interdisciplinary fields that recently emerged between these areas. On the other hand, as the boundary separating fundamental research and applied technology is more and more dissolving, this series is particularly open to trans-disciplinary topics between fundamental science and engineering.

Indexed by EI-Compendex, SCOPUS and Springerlink.

Wojciech Czekała

Biological Treatment of Waste and By-Products from Food Industry

 Springer

Wojciech Czekała (iD)
Faculty of Environmental and Mechanical
Engineering
Poznań University of Life Sciences
Poznań, Poland

ISSN 2191-530X   ISSN 2191-5318 (electronic)
SpringerBriefs in Applied Sciences and Technology
ISBN 978-3-031-47486-6   ISBN 978-3-031-47487-3 (eBook)
https://doi.org/10.1007/978-3-031-47487-3

This Springer imprint is published by the registered company Springer Nature Switzerland AG
The registered company address is: Gewerbestrasse 11, 6330 Cham, Switzerland

Paper in this product is recyclable.

To my mother and father, who never stopped believing in me.
To my wife for understanding me better than everyone.
To my sons, who fill my heart with joy each and every day.

Contents

Chapter 1
Food Industry as a Source of Waste and By-Products

1.1 Introduction

In recent years, there has been a growing interest in environmental topics (Cortat Simonetti Goncalves 2023). As part of the activities undertaken, issues such as waste management are gaining importance (Moustakas and Loizidou 2023). Due to economic growth and improved living conditions of people on earth, the amount of waste generated is increasing (Czekała et al. 2023a). Waste management regulations, especially in the context of the construction of new installations, are becoming stricter. This is particularly noticeable in European Union countries (Marques and Teixeira 2022). The same is true for the management of the waste generated. The waste hierarchy clearly indicates that recycling should be one of the preferred waste management methods.

Sustainability and the circular economy are an issue that both the economy and the municipal sectors are increasingly paying attention to (Fischer et al. 2023). Sustainability is related to take actions according to which the needs of the present generation can be met without diminishing the chances of future generations to meet them (Rodríguez-Antón et al. 2022). Initially, sustainable development referred primarily to actions related to the impact of humans and the economy on the natural environment. However, this was insufficient and the approach to the topic changed in subsequent years. In addition to the reduction of negative environmental impacts, additional aspects began to gain in importance, which included economic growth together with social progress. Another aspect addressed in the context of sustainable development will be education (Veidemane 2022). As a result of activities, sustainability should be addressed in every aspect concerning the environmental, social and economic elements (Jeronen 2022). For this reason, the emphasis is on integrated action involving all parts of society.

One of the activities related to sustainability is the circular economy (Ramirez-Corredores et al. 2023). Circular economy is a model involving production and consumption that is environmentally friendly. It includes using all kinds of products

W. Czekała, *Biological Treatment of Waste and By-Products from Food Industry*,
SpringerBriefs in Applied Sciences and Technology,
https://doi.org/10.1007/978-3-031-47487-3_1

and materials in such a way that they stay in circulation as long as possible (Alhawari et al. 2021). This is directly related to the life cycle of products, which should be extended. As a result of the measures taken, the amount of waste generated that should be managed will be reduced. By using this waste in recycling and recovery processes, the part of the raw materials is recycled, which will reduce the amount of raw materials that must be obtained and processed to make new products. Circular economy should be used instead of linear economy, a system that is characterised by the product discard at the end of product life (Michelini et al. 2017).

The food industry is a very important part of any country's economy (Rejeb et al. 2022). It is a key aspect related to the food security of a country and its people. For many countries, it is important not only for their operations but also for the ability to export certain types of food to other countries (Elneel 2023). The agri-food industry encompasses a very broad and diverse production area which includes:

- fruit and vegetable industry,
- grain industry,
- bakery and confectionery industry,
- meat industry,
- fishing industry,
- dairy industry,
- oil and fat industry,
- soft drink industry,
- alcoholic beverage industry,
- feed industry.

The aim of processes in the agri-food industry is to produce finished goods. However, in addition to the main product, various types of residues are generated, and depending on their characteristics, these may be referred to as by-product or waste (Zhu et al. 2020). Every agri-food industry generates waste as an integral part of the processes involved in food production. Food-related waste is generated in many places including production facilities, logistics centres, markets, shops, and households (Khedkar and Singh 2018). Research on circular economy in the food industry is being conducted worldwide (Jurgilevich et al. 2016). They have specific properties that are characteristic of the substrates from which they are derived and because of the products that arise from their processing. A common feature of most waste in this group is that they are biodegradable under aerobic and anaerobic conditions including the influence of microorganisms. The organic matter content is very high and in many cases exceeds 90% on a dry matter basis. Thanks to these properties, they can be processed in biological waste treatment processes.

1.2 Food Industry Sectors

The food industry is one of the most important and fastest-growing industries in many countries of the world (Fukase and Martin 2020). The food industry is a specific but very important industry for each country. This is due to the fact that access to food is one of the basic elements necessary for every human being to survive. The food industry encompasses many sectors with the common goal of providing quality food products (McCarthy 2023). This applies to all stages of the production chain. This imposes an obligation on companies to ensure the quality of their products and their wide range of offerings in order to meet the growing expectations of customers. A high technical and technological level and appropriate organisation of work, especially production, makes it possible to obtain valuable products, which will be used by the inhabitants of a country and can also be exported, influencing the country's gross domestic product. Therefore, innovation is needed to positively influence the processing of food products (Alawamleh et al. 2022). Within the food industry, sectors can be separated, and the most important sectors are summarised in Table 1.1.

The objects of the food industry are living organisms or their parts. They are characterised by high instability (susceptibility to decomposition) (Mu et al. 2023), which also applies to semi-finished and finished goods, although their shelf life is often higher. The food industry is growing rapidly in many countries around the world, such as India (Sharma et al. 2023). The high dispersion and fragmentation of farms in many countries, e.g. Poland, may be responsible for the distribution of individual processing plants. This is primarily related to the raw material base. The individual sectors are briefly discussed below.

Fruit and Vegetable Industry It includes activities related to the primary and further processing of fruit and vegetables into preparations such as juices and

Table 1.1 Characteristics of selected sectors including the food industry

Sector	Characteristics of the sector
Fruit and vegetable industry	Processing and preserving fruit and vegetables
Grain industry	Manufacture of grain mill products, starches and starch products
Bakery and confectionery industry	Manufacture of bakery and farinaceous products
Meat industry	Processing and preserving of meat and meat goods production
Fishing industry	Processing and preserving fish, crustaceans and molluscs
Dairy industry	Manufacture of dairy products
Oil and fat industry	Production of oils and fats vegetable and animal origin
Soft drink industry	Production of non-alcoholic beverages
Alcoholic beverage industry	Production of alcoholic beverages
Feed industry	Production of fodder and animal feed

beverages, frozen foods, jams and preserves, fruit and vegetable pickles, vegetable preserves and dried foods, among others. Many fruit and vegetables are intended for direct consumption.

Grain Industry Includes activities related to the trading of cereals (purchase and storage), the milling of cereals (for food and feed purposes) and the production of cereal products with a long expiration date, e.g. pasta. In addition, businesses selling cereals and those providing services including cleaning, drying and storage of cereals should be included.

Bakery and Confectionery Industry Includes activities related to the making and baking of bread and confectionery products, both in craft and industrial bakeries and confectioneries.

Meat Industry It includes activities related to the processing of meat raw materials, such as animals slaughtering, meat cutting into parts and the production of meat products, offal and meat preserves. An additional area of the meat industry is the production of ready-made meat dishes including portioning, slicing and packaging.

Fishing Industry Includes activities related to the processing of marine and inland catches into finished goods, including frozen foods. It deals, among other things, with the portioning, filleting, marinating and smoking of fish and the packaging of fish into canned products.

Dairy Industry Includes activities related to the purchase and processing of milk into dairy products, e.g. drinking milk, dairy drinks, butter, cream, cheese, milk powder and ice cream.

Oil and Fat Industry Includes activities related to the processing of oilseed crops and the production of goods such as margarines, vegetable fats and edible oils.

Soft Drink Industry Includes activities related to the production of non-alcoholic beverages.

Alcoholic Beverage Industry Includes activities related to the production of alcoholic beverages. The alcoholic beverage industry is most commonly divided into the brewing industry, the wine industry and the spirits industry.

Feed Industry Includes activities related to the production of animal feed and food.

1.3   Food Industry as a Source of Waste and By-Products

The topics of food production, food consumption and food waste are closely related. This is due to the functioning of the whole cycle starting with the acquisition of raw materials and ending with the management of the waste. Therefore, all processes should be considered together. Given that food is consumed by everyone, this is a problem that affects every country in the world. However, it is important to emphasise that the specific aspects or key problems related to food supply and food waste may differ significantly in different places around the world. For example, developed countries may have a problem with food overproduction and food waste, while other countries may benefit significantly from imported food. Regardless of these dependencies, however, waste and by-products are produced everywhere on earth. While there is overproduction of food in some countries, there is starvation in others. The simultaneous coexistence of food loss and waste and hunger, undernourishment and malnutrition represents a failure of contemporary food systems (von Braun et al. 2023).

The food industry is characterised by a large diversity of factories, not only by type of production, but also by scale (Czekała et al. 2017). For example, in Poland there are several brewery groups whose production covers a significant part of the market, as well as regional, craft or restaurant breweries. It should be noted that a significant number of all factories are dispersed. Therefore, waste and by-products are also available in many locations. Processing plants are often placed in locations directly linked to the raw material base or markets, thus reducing transport costs. Since waste from the food industry is generated in every manufacturing plant, every day, in huge quantities, the need arises to manage it on an ongoing basis (Kaur et al. 2023). It is worth mentioning that waste management issues affect areas locally, but with the accumulation of problems or threats, it can be regional or even global.

In order to achieve efficient waste management, regulations and guidelines on waste handling are necessary. An extremely important issue in the context of waste management is the waste hierarchy. This hierarchy indicates the preferred order of actions in relation to waste (López-Portillo et al. 2021) and was introduced into the national legal system through the regulations contained in Directive 2008/98/EC of the European Parliament and of the Council of 19 November 2008 on waste and repealing certain Directives (Text with EEA relevance) (2008). The waste hierarchy is presented in the form of an inverted pyramid in Fig. 1.1.

Fig. 1.1 Waste hierarchy (based on Directive 2008)

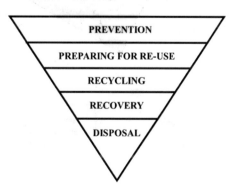

This hierarchy includes five elements:

(1) Prevention—"measures taken before a substance, material or product has become waste, that reduce: (a) the quantity of waste, including through the re-use of products or the extension of the life span of products; (b) the adverse impacts of the generated waste on the environment and human health; or (c) the content of harmful substances in materials and products".

(2) Preparing for re-use—"checking, cleaning or repairing recovery operations, by which products or components of products that have become waste are prepared so that they can be re-used without any other pre-processing".

(3) Recycling—"any recovery operation by which waste materials are reprocessed into products, materials or substances whether for the original or other purposes. It includes the reprocessing of organic material but does not include energy recovery and the reprocessing into materials that are to be used as fuels or for backfilling operations".

(4) Recovery—"any operation the principal result of which is waste serving a useful purpose by replacing other materials which would otherwise have been used to fulfil a particular function, or waste being prepared to fulfil that function, in the plant or in the wider economy".

(5) Disposal—"any operation which is not recovery even where the operation has as a secondary consequence the reclamation of substances or energy" (Directive 2008).

The waste hierarchy indicates waste management possibilities. This is directly related to their environmental impact. Therefore, prevention will be the most preferred direction and disposal the least preferred. Waste management is a very broad and complex issue covering the collection, transport, recovery and disposal of waste, including the supervision of such operations and the after-care of disposal sites and including actions taken as a dealer or broker (Directive 2008). The waste management process in simplified graphical form is presented in Fig. 1.2.

The processing of agricultural crops into food products is a complex issue. Agricultural production and the food industry are responsible for generating a very large amount of waste and by-products. By-products and waste have an impact on the environment, the economy and society. By-products and inedible parts can represent

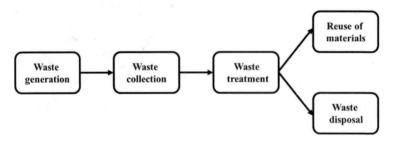

Fig. 1.2 Management of food industry waste

up to more than 50% of the weight of the substrates, which must be considered a very large amount.

The first stage is the production of waste within the company. Every step in the processing of different products is associated with the generation of waste. If several criteria are met, it is possible for a given raw material not to be classified as waste, but as a by-product. According to Directive (2008) "a substance or object, resulting from a production process, the primary aim of which is not the production of that item, may be regarded as not being waste referred to in point (1) of Article 3 but as being a by-product only if the following conditions are met:

(a) further use of the substance or object is certain;
(b) the substance or object can be used directly without any further processing other than normal industrial practice;
(c) the substance or object is produced as an integral part of a production process; and
(d) further use is lawful, i.e. the substance or object fulfils all relevant product, environmental and health protection requirements for the specific use and will not lead to overall adverse environmental or human health impacts".

The waste and by-product generated must be collected at the place of generation. Due to the wide variety, the method of collection will also vary. Waste should be stored in a way that is appropriate to its physical and chemical properties. They are most often collected using containers, tanks, bins, bags or heaps in the case of waste where there is no risk of soil, ground and water contamination. Waste and by-product should be collected in places where the storage capacity will be adapted to the mass of waste generated at a given time. The time of collection should also be taken into account, as the more often the waste is collected, the smaller the storage capacity may be. Consideration should still be given to whether the waste is generated seasonally, such as beet pulp, or in similar quantities throughout the year, such as waste from the meat industry (Czekała 2018). Until the waste is collected, it should be stored in a way that prevents it from spreading beyond its intended location on the site, let alone outside the perimeter. The waste is then taken by specialised companies to the sites where it will be treated or stored. In the case of the substrates in question, most of them can be successfully used for energy or fertiliser production, among other things. Only some of those that are not suitable for recycling or recovery will be stored.

As Garcia-Garcia et al. (2017) point out, an extremely important element in the context of food waste management is their classification based on their specific characteristics. Similarly, this is the case for waste and by-products from food industry. In the following, a list of selected waste and by-products generated in the different sectors of the agri-food industry is presented (Table 1.2). The list prepared shows the considerable amount and variety of raw materials discussed, which will have an impact on their use. Table 1.2 mainly lists those waste that can be used in biological waste treatment processes, i.e. composting or anaerobic digestion.

The consumption of fruit and vegetables is particularly important for the health of the population. These products should have a special role in nutrition (Bracale

Table 1.2 List of selected waste and by-products generated by the different sectors of the agri-food industry

Sector	Sector characteristics
Fruit and vegetable industry	Fruit and vegetables not for consumption Fruit and vegetable pomace Potato pulp Seeds Peelings Grain Sediments Leaves Roots Stems
Grain industry	Unmanaged crops Contaminated grain Cereal straw Maize stalks Husks
Bakery and confectionery industry	Stale bread Bread rolls Cookies Confectionery waste Creamy and non-creamy bakery waste Beet pulp
Meat industry	Animal tissue Digestive contents Intestines Bones Blood Skin Fats Hair Feathers Horns Hooves High risk waste
Fishing industry	Animal tissue Waste water
Dairy industry	Residues of milk, cottage cheese and cheese Whey Permeate Buttermilk Waste water
Oil and fat industry	Pomace after extraction Slurries
Soft drink industry	Fruit and vegetables not for consumption Fruit and vegetable pomace

(continued)

Table 1.2 (continued)

Sector	Sector characteristics
Alcoholic beverage industry	Brewers' malt Distillers' stillage Pomace Sludge Effluents
Feed industry	Unused parts of raw materials

Fig. 1.3 Fruit that has not been sold as an example of biogas feedstock from food industry

et al. 2020; Brouwer et al. 2021). Jun et al. (2022) based on Li et al. (2021) indicated that around a quarter of all fruits and vegetables in the supply chain is lost due to wastage. The quantity and variety of waste in this group will be particularly high (Czekała et al. 2016). The predominant raw materials in this group include unfit fruits and vegetables, fruit and vegetable pomace and inedible plant parts such as seeds, peels or skins. It should be noted that in addition to biological methods of processing these substrates, other uses are also becoming popular. Fruits, pit peel and pomace, can be used for chemical and biofuels substances production via benign conversion procedures (Ehsana and Parsimehr 2020). Research on the use of fruit pits as aggregate in concreto was conducted by Grubeša et al., (2022). The results indicated that replacing the aggregate by fruit pits decreased the density, influenced the compressive strength negatively and reduced thermal conductivity of concreto. Figure 1.3 shows the fruits that will be feedstock for biogas production.

Cereals are a group of crops used for consumption, fodder and industrial purposes. The grain industry is a sector that is developed in almost every country. This is due to the fact that cereals are one of the basic elements in human nutrition. Cereals are an essential raw material for many other industries besides milling, such as brewing, distilling and pharmaceuticals. According to Erenstein et al. (2022a), wheat has played a crucial role in global food security, and it is the most widely cultivated crop in the world (cultivated on 217 million ha annually). Maize is already the leading cereal in terms of production. Maize is set to become the most widely grown and traded crop in the coming decade Erenstein et al. (2022b). The authors point out that maize has played an increasing role in agri-food systems globally. The waste generated in this sector is characterised by a significantly higher dry matter content than waste from other branches. For example, Van Hung et al. (2020) presented dry matter values of rice straw published by selected authors. These values ranged from 90.6 to 96.3%. It is this parameter that will largely determine the directions of use of the biomass in question. Straw can be used in many sectors, e.g. as bedding for animals (Ferraz et al. 2020). Another direction of straw use will be in the production of solid biofuels, as discussed in the work of Kizuka et al. (2021), among others.

Waste from the bakery and confectionery industry comprises mainly shop returns and food that do not meet the criteria. These include stale bread, bread rolls, cookies, confectionery waste, creamy and non-creamy bakery waste. Due to their nutrient content and high organic matter content, waste from this group should be used primarily in anaerobic digestion. As in the case of the grain industry, bakery waste with a high dry matter content can also be used to produce solid biofuels, as investigated by Dołżyńska et al. (2019), among others.

The management of waste from the meat industry is particularly important. Waste from the meat industry is specific and requires special procedures for its management (Staroń et al. 2017). Waste from slaughterhouses can be a source of pathogens that are dangerous to humans and animals. Effective and safe disposal methods are needed to reduce the spread of diseases following animal slaughtering (Mozhiarasi and Natarajan 2022). Therefore, methods using high temperatures are often used to treat waste from meat industry (Lee et al. 2021). The meat industry includes companies involved in animal slaughter, cutting up carcasses, trimming meat and producing meat products. Therefore, the variety of waste in this sector will be very high, relative to other sectors. The meat industry generates significant amounts of waste including waste animal tissue, bones, blood and skin. Fish waste includes for example heads, viscera, chitinous material and wastewater (Ben Rebah and Miled 2013). As with meat industry waste, the management of fishing industry waste and by-product can also be an environmental and health issue.

According to Mazurkiewicz (2022), dairy farms are a major contributor to the total life cycle greenhouse gas emissions of milk and other dairy products. The main waste from the dairy industry is whey and buttermilk. Buttermilk is a by-product of processing cream into butter. Whey, on the other hand, is a by-product from the production of products such as cheese and cottage cheese. The properties of these raw materials will be closely related to the parameters of the milk used in the processes. What they have in common, however, will be a high water content, which is how they differ from, for example, waste from the grain industry. For this reason, waste and by-products from the dairy industry should be used primarily in anaerobic digestion. This is particularly important because chemical oxygen demand is reduced as a result of this process (Chen et al. 2022). It should be noted that the dairy sector requires large volumes of water, which generates significant amounts of wastewater (Raghunath et al. 2016). For this reason, wastewater treatment plants are often operated next to production facility.

Fats and oils are essential components of the human diet. They are the main source of energy and carrier of fat-soluble vitamins (A, D, E, K). Fats are a valuable raw material not only in the food and feed industry, but also in the cosmetic or pharmaceutical industry. It should be emphasised that oil and fats and the waste associated with their production are a substrate with high energy potential. Among the main waste in this branch are pomace and post-extraction meal. The use of fats and oils in the production of biofuels is also increasing every year. This also applies to less common plant species. For example, Yadav et al. (2017) conducted research on the production of biodiesel from oleander (*Thevetia peruviana*). An interesting direction is the use of algae. Microalgae and macroalgae have a potential for several industrial applications. As reported by Mahmood et al. (2023), developing of algal cultivation for a biofuel industry can provide a lot of socio-economic advantages.

The next sectors are soft drink and alcoholic beverage. The term soft drink, due to its wide scope, can be interpreted in different ways. As reported by Varnam and Sutherland (1994), the widest interpretation is that the term encompasses all non-alcoholic drinks, including non-alcoholic beer and wine and water and excluding products such as tea, coffee and milk-based drinks. Over the past century, soft drinks have changed significantly. From being a local pharmacy product to worldwide industry (Tahmassebi and BaniHani 2020), the waste and products generated in this sector will be very similar to those generated in the fruit and vegetable industry and mainly include unfit for consumption fruits and vegetables and fruit and vegetable pomace.

Alcoholic beverage consumption has a long history. Alcoholic beverage consumption plays an important role in many countries. In some regions, alcohol consumption is considered part of the diet. This is despite the fact that it can have harmful effects on the health of those who consume it (Skinner et al. 2023). Within the alcoholic beverage group, a distinction can be made between beers, wines and spirits. Waste generated in the production of alcoholic beverages includes brewer's malt and distillers' decoction. It should also be noted that this sector of the food industry consumes significant amounts of water.

The feed industry involves the production of feed for animal consumption. This includes companion, wild, exotic species and fish (Boyd 2020). Although this industry concerns animals, it also indirectly affects many aspects of human life. This is because the feed industry supports the production of animal products destined for the human food chain. As the cost of purchasing feed is the largest of the expenses associated with animal husbandry, the role of this industry in animal production is significant. The use of agro-industrial waste for livestock feed is an extremely important issue in the context of waste management and the reduction of production costs, which is being studied by many researchers. For example, research on Mexico is presented in the work of Quintero-Herrera et al. (2023). By some researchers, the feed industry is treated separately and not as a food industry sector.

The food industry is diverse and encompasses many different sectors (Gaitán-Cremaschi et al. 2019). Each operates in its own distinctive way, and the production process itself and the waste and by-products generated vary widely. The volume of production also varies, as within each sector, there can be either local plants or large enterprises employing hundreds or even thousands of workers. From the information presented above, the variety of waste and by-products is large (Rashwan et al. 2023). Here, one can find both liquid waste, such as distillery stock or dairy industry effluent, and solid waste such as straw or pomace. The diversity, which can be a problem, is often an advantage in terms of management options (Czekała et al. 2023b). Depending on the processing, suitable raw materials can be selected. Of particular note is the fact that most of the waste and by-products from food industry can be managed in more than one way.

In some countries, a significant problem related to the management of waste and by-products from food industry is their record-keeping in terms of quantities as well as types of waste generated (Malik et al. 2022). This is particularly true for small and medium-sized enterprises. Some of the waste generated may not be registered and passed on to, for example, farmers use as animal feed. There are also cases of illegal management such as dumping waste on fields or pouring sewage into rivers and lakes. As a result, the amount and type of waste generated is underreported. A situation where there is a lack of data on the current status can adversely affect the planning and selection of management methods and sites. However, in many countries the system is becoming increasingly effective, although not perfect. The implementation in Poland of the integrated BDO system—Database on Products and Packaging and Waste Management (Banaszkiewicz et al. 2022)—is an example of the measures taken in this respect. The benefits of this tool include optimising the waste management reporting process, increasing control over the national waste management and ensuring monitoring of the flow of waste streams. These measures will reduce irregularities in the waste management sector, although they will probably not completely rule them out.

The management of such large quantities of diverse biodegradable substrates is a huge challenge (Frankowski and Czekała 2023). Regardless of the measures, care should be taken to ensure a complete inventory of the waste generation sites and the amount of waste. Carrying out an inventory in accordance with the actual state of affairs will make it possible to propose appropriate practices and select waste management methods. It will also make it possible to avoid a situation where there may be facilities in some regions that have problems obtaining feedstock, while other regions will have no facilities or insufficient capacity. Due to their enormous quantity and diversity, many methods are used to manage them. It should also be noted that they are processed into valuable products in many different sectors. These include the food sector, the energy sector and the fertiliser sector.

The vast majority of waste generated in the agri-food industry is classified as organic waste. An organic matter content of more than 50% (sometimes even more than 90%) determines their manageability. The various sectors of the agri-food industry mainly generate organic waste and by-products of plant and animal origin (Bertocci and Mannino 2022). The substrates in question are characterised by properties that allow them to be used for feed, fertiliser and energy purposes.

Among the ways to manage waste and by-products of the agri-food industry are:

- human consumption and use (zero waste idea),
- use in animal nutrition,
- industrial use of individual components,
- land reclamation,
- fertilisation,
- composting,
- anaerobic digestion,
- thermal processes.

Increasingly, scientists are researching alternative directions for waste management. As noted by Dey et al. (2021), most by-products are highly nutritious and can be excellent low-cost sources of dietary fibre, proteins and bioactive compounds such as polyphenols, antioxidants and vitamins. Another direction of their development may be green technology to produce humic. Research on this topic was conducted by Kanmaz (2019), using three food by-products as flaxseed meal, lemon peel and mandarin peel. Another interesting direction may be the use of waste from the food industry for the production of biotextiles. Waste from one industry can become a new product for another industry, in accordance with the principle of interdisciplinarity. The paper mentions, among others, on the possibility of use of bacterial cellulose from the probiotic drinks (from kombucha), for the manufacture of biotextiles for fashion industry.

Choosing a method of waste and by-products management related to food production is a complex task. When choosing the direction of development, the following should be taken into account: legal conditions, environmental impact, financial capabilities of the waste owner or holder and the possibility of producing additional products such as biogas (Czekała 2022). The specificity of a specific production plant and local conditions should also be taken into account.

1.4 Conclusion

The food industry plays an important role in the economy of each country. Its most important function is to ensure food security. The impact of the food industry should be analysed in environmental, economic and social terms. The manufactured food products can be sold on the domestic market, imported or exported. Due to the significant diversity of individual branches of the food industry, waste and by-product from the food industry will also be diverse.

Due to the fact that the food industry is very diverse, waste and by-product from the food industry also are a large group of substrates. It is characterised by great diversity, although some common features such as high organic matter content and susceptibility to decomposition are common. The high availability of these substrates around the world creates the need for their responsible management.

References

M. Alawamleh, M. Al-Hussaini, L. Bani Ismail, Open innovation in the food industry: trends and barriers—A case of the Jordanian food industry. J Glob Entrepr Res **12**, 279–290 (2022). https://doi.org/10.1007/s40497-022-00312-6

O. Alhawari, U. Awan, M.K.S. Bhutta, M.A. Ülkü, Insights from circular economy literature: a review of extant definitions and unravelling paths to future research. Sustainability **13**, 859 (2021). https://doi.org/10.3390/su13020859

K. Banaszkiewicz, I. Pasiecznik, W. Cieżak, E. den Boer, Household E-waste management: a case study of Wroclaw, Poland. Sustainability **14**, 11753 (2022). https://doi.org/10.3390/su1 41811753

F. Ben Rebah, N. Miled, Fish processing wastes for microbial enzyme production: a review. 3 Biotech **3**, 255–265 (2013). https://doi.org/10.1007/s13205-012-0099-8

F. Bertocci, G. Mannino, Can agri-food waste be a sustainable alternative in aquaculture? A bibliometric and meta-analytic study on growth performance, innate immune system, and antioxidant defenses. Foods **11**, 1861 (2022). https://doi.org/10.3390/foods11131861

J. Boyd, Feed industry, in *Encyclopedia of Animal Cognition and Behavior*, ed. by J. Vonk, T. Shackelford (Springer, Cham, 2020). https://doi.org/10.1007/978-3-319-47829-6_222-1

R. Bracale, C.M. Vaccaro, V. Coletta, C. Cricelli, F.C. Gamaleri, F. Parazzini, M. Carruba, Nutrition behaviour and compliance with the Mediterranean diet pyramid recommendations: an Italian survey-based study. Eat. Weight Disord. **25**, 1789–1798 (2020). https://doi.org/10.1007/s40519-019-00807-4

J. von Braun, M.S. Sorondo, R. Steiner, Reduction of food loss and waste: the challenges and conclusions for actions, in *Science and Innovations for Food Systems Transformation*, ed. by von J. Braun, K. Afsana, L.O. Fresco, M.H.A. Hassan (Springer, Cham, 2023). https://doi.org/10.1007/978-3-031-15703-5_31

I.D. Brouwer, M.J. van Liere, A. de Brauw, P. Dominguez-Salas, A. Herforth, G. Kennedy, C. Lachat, E.B. Omosa, E.F. Talsma, S. Vandevijvere, J. Fanzo, M. Ruel, Reverse thinking: taking a healthy diet perspective towards food systems transformations. Food Sec. **13**, 1497–1523 (2021). https://doi.org/10.1007/s12571-021-01204-5

L. Chen, J. Qin, Q. Zhao, Z. Ye, Treatment of dairy wastewater by immobilized microbial technology using polyurethane foam as carrier. Bioresour. Technol. **347**, 126430 (2022). https://doi.org/10.1016/j.biortech.2021.126430

L. Cortat Simonetti Goncalves, EU environmental law in face of 2022 challenges: a Prufrock-like confession during times when one should dare? ERA Forum **24**, 1–7 (2023). https://doi.org/10.1007/s12027-023-00750-z

W. Czekała, J. Drozdowski, P. Łabiak, Modern technologies for waste management: a review. Appl. Sci. **13**, 8847 (2023a). https://doi.org/10.3390/app13158847

W. Czekała J. Pulka, T. Jasiński, P. Szewczyk, W. Bojarski, J. Jasiński, Waste as substrates for agricultural biogas plants: a case study from Poland. J. Water Land Dev. **56**(I–III), 1–6 (2023b). https://doi.org/10.24425/jwld.2023.143743

W. Czekała, A. Smurzyńska, M. Cieślik, P. Boniecki, K. Kozłowski, Biogas efficiency of selected fresh fruit covered by the Russian embargo, in *Energy and Clean Technologies Conference Proceedings, SGEM 2016*, vol. III (2016), pp. 227–233. https://doi.org/10.5593/sgem2016HB43

W. Czekała, K. Gawrych, A. Smurzyńska, J. Mazurkiewicz, A. Pawlisiak, D. Chełkowiski, M. Brzoski, The possibility of functioning micro-scale biogas plant in selected farm. J. Water Land Dev. **35**(X–XII), 19–25 (2017). https://doi.org/10.1515/jwld-2017-0064

W. Czekała, Agricultural biogas plants as a chance for the development of the agri-food sector. J. Ecol. Eng. **19**(2), 179–183 (2018). https://doi.org/10.12911/22998993/83563

W. Czekała, Biogas as a sustainable and renewable energy source, in *Clean Fuels for Mobility. Energy, Environment, and Sustainability*, ed. by G. Di Blasio, A.K. Agarwal, G. Belgiorno, P.C. Shukla (Springer, Singapore, 2022). https://doi.org/10.1007/978-981-16-8747-1_10

D. Dey, J.K. Richter, P. Ek, B.-J. Gu, G.M. Ganjyal, Utilization of food processing by-products in extrusion processing: a review. Front. Sustain. Food Syst. **4**, 603751 (2021). https://doi.org/10.3389/fsufs.2020.603751

Directive 2008/98/EC of the European Parliament and of the Council of 19 November 2008 on waste and repealing certain Directives (Text with EEA relevance)—Consolidated text. Accessed 31 July 2023

M. Dołżyńska, S. Obidziński, M. Kowczyk-Sadowy, M. Krasowska, K. Karczewski, D. Jóźwiak, R. Buczyński, pressure agglomeration process of bakery industry waste. Proceedings **16**, 37 (2019). https://doi.org/10.3390/proceedings2019016037

A. Ehsana, H. Parsimehr, Electrochemical energy storage electrodes from fruit biochar. Adv. Colloid Interface Sci. **284**, 102263 (2020). https://doi.org/10.1016/j.cis.2020.102263

F.A. Elneel, Food exports and its contribution to the sudan economic development in the light of covid-19 pandemic: evidence from simultaneous equation model (1974–2019), in *From Industry 4.0 to Industry 5.0. Studies in Systems, Decision and Control*, vol. 470, ed. by A. Hamdan, A. Harraf, A. Buallay, P. Arora, H. Alsabatin (Springer, Cham, 2023). https://doi.org/10.1007/978-3-031-28314-7_25

O. Erenstein, M. Jaleta, K.A. Mottaleb, K. Sonder, J. Donovan H.J. Braun, Global trends in wheat production, consumption and trade, in *Wheat Improvement*, ed. by M.P. Reynolds, H.J. Braun (Springer, Cham, 2022a). https://doi.org/10.1007/978-3-030-90673-3_4

O. Erenstein, M. Jaleta, K. Sonder, K. Mottaleb, B.M. Prasanna, Global maize production, consumption and trade: trends and R&D implications. Food Sec. **14**, 1295–1319 (2022b). https://doi.org/10.1007/s12571-022-01288-7

P.F.P. Ferraz, G.A. Silva Ferraz, L. Leso, M. Klopčič, M. Barbari, G. Rossi, Properties of conventional and alternative bedding materials for dairy cattle. J. Dairy Sci. **103**(9), 8661–8674 (2020). https://doi.org/10.3168/jds.2020-18318

M. Fischer, D. Foord, J. Frecè, K. Hillebrand, I. Kissling-Näf, R. Meili, M. Peskova, D. Risi, R. Schmidpeter, T. Stucki, The concept of sustainable development, in *Sustainable Business. SpringerBriefs in Business* (Springer, Cham, 2023). https://doi.org/10.1007/978-3-031-25397-3_2

J. Frankowski, W. Czekała, Agricultural plant residues as potential co-substrates for biogas production. Energies **16**(11), 4396 (2023). https://doi.org/10.3390/en16114396

E. Fukase, W. Martin, Economic growth, convergence, and world food demand and supply. World Dev. **132**, 104954 (2020). https://doi.org/10.1016/j.worlddev.2020.104954

D. Gaitán-Cremaschi, L. Klerkx, J. Duncan, J.H. Trienekens, C. Huenchuleo, S. Dogliotti, M.E. Contesse W.A.H. Rossing, Characterizing diversity of food systems in view of sustainability transitions. A review. Agron. Sustain. Dev. **39**(1), (2019) https://doi.org/10.1007/s13593-018-0550-2

G. Garcia-Garcia, E. Woolley, S. Rahimifard, J. Colwill, R. White, L. Needham, A methodology for sustainable management of food waste. Waste Biomass Valor **8**, 2209–2227 (2017). https://doi.org/10.1007/s12649-016-9720-0

I.N. Grubeša, B. Marković, M.H. Nyarko, H. Krstić, J. Brdarić, N. Filipović, I.S.Á. Kukovecz, Potential of fruit pits as aggregate in concreto. Constr. Build. Mater. **345**, 128366 (2022). https://doi.org/10.1016/j.conbuildmat.2022.128366

E. Jeronen, Education for sustainable development, in *Encyclopedia of Sustainable Management*, ed. by S. Idowu, R. Schmidpeter, N. Capaldi, L. Zu, M. Del Baldo, R. Abreu (Springer, Cham, 2022). https://doi.org/10.1007/978-3-030-02006-4_351-1

Y. Jun, W. Yifan, W. Qiongyin, Z. Shuo, W. Meizhen, F. Huajun, J. Jun, Q. Xiaopeng, Z. Yanfeng, C. Ting, Generation of fruit and vegetable wastes in the farmers' market and its influencing factors: a case study from Hangzhou. China. Waste Manage. **154**, 331–339 (2022). https://doi.org/10.1016/j.wasman.2022.10.023

A. Jurgilevich, T. Birge, J. Kentala-Lehtonen, K. Korhonen-Kurki, J. Pietikäinen, L. Saikku, H. Schösler, Transition towards circular economy in the food system. Sustainability **8**, 69 (2016). https://doi.org/10.3390/su8010069

E.O. Kanmaz, Humic acid formation during subcritical water extraction of food by-products using accelerated solvent extractor. Food Bioprod. Process. **115**, 118–125 (2019). https://doi.org/10.1016/j.fbp.2019.03.008

M. Kaur, A.K. Singh, A. Singh, Bioconversion of food industry waste to value added products: current technological trends and prospects. Food Biosci. **55**, 102935 (2023). https://doi.org/10.1016/j.fbio.2023.102935

R. Khedkar, K. Singh, Food industry waste: a panacea or pollution hazard? in *Paradigms in Pollution Prevention. SpringerBriefs in Environmental Science*, ed. by T. Jindal (Springer, Cham, 2018). https://doi.org/10.1007/978-3-319-58415-7_3

R. Kizuka, K. Ishii, S. Ochiai, M. Sato, A. Yamada, K. Nishimiya, Improvement of biomass fuel properties for rice straw pellets using torrefaction and mixing with wood chips. Waste Biomass Valor **12**, 3417–3429 (2021). https://doi.org/10.1007/s12649-020-01234-8

J. Lee, S. Cho, D. Kim, J. Ryu, K. Lee, H. Chung, K.Y. Park, Conversion of slaughterhouse wastes to solid fuel using hydrothermal carbonization. Energies **14**, 1768 (2021). https://doi.org/10.3390/en14061768

X. Li, X. Liu, S. Lu, G. Cheng, Y. Hu, J. Liu, Z. Dou, S. Cheng, G. Liu, China's food loss and waste embodies increasing environmental impacts. Nat. Food **2**, 519–528 (2021). https://doi.org/10.1038/s43016-021-00317-6

M.P. López-Portillo, G. Martínez-Jiménez, E. Ropero-Moriones, M.S. Saavedra-Serrano, Waste treatments in the European Union: a comparative analysis across its member states. Helyion **7**(12), e08645 (2021). https://doi.org/10.1016/j.heliyon.2021.e08645

T. Mahmood, N. Hussain, A. Shahbaz, S.I. Mulla, H.M.N. Iqbal, M. Bilal, Sustainable production of biofuels from the algae-derived biomass. Bioprocess Biosyst. Eng. **46**, 1077–1097 (2023). https://doi.org/10.1007/s00449-022-02796-8

M. Malik, S. Sharma, M. Uddin, C.L. Chen, C.-M. Wu, C.S. SoniP, Waste classification for sustainable development using image recognition with deep learning neural network models. Sustainability **14**, 7222 (2022). https://doi.org/10.3390/su14127222

A.C. Marques, N.M. Teixeira, Assessment of municipal waste in a circular economy: do European Union countries share identical performance? Clean Waste Syst **3**, 100034 (2022). https://doi.org/10.1016/j.clwas.2022.100034

J. Mazurkiewicz, Energy and economic balance between manure stored and used as a substrate for biogas production. Energies **15**, 413 (2022). https://doi.org/10.3390/en15020413

J. McCarthy, Influence of the food industry: the food environment and nutrition policy, in *Nutritional Health. Nutrition and Health*, ed. by, N.J. Temple, T. Wilson, D.R. Jacobs, Jr., G.A. Bray (Humana, Cham, 2023). https://doi.org/10.1007/978-3-031-24663-0_30

G. Michelini, R.N. Moraes, R.N. Cunha, J.M.H. Costa, A.R. Ometto, From linear to circular economy: PSS conducting the transition. Procedia CIRP **64**, 2–6 (2017). https://doi.org/10.1016/j.procir.2017.03.012

K. Moustakas, M. Loizidou, Effective waste management with emphasis on circular economy. Environ. Sci. Pollut. Res. **30**, 8540–8547 (2023). https://doi.org/10.1007/s11356-022-24670-6

V. Mozhiarasi, T.S. Natarajan, Slaughterhouse and poultry wastes: management practices, feedstocks for renewable energy production, and recovery of value added products. Biomass Conv. Bioref. (2022). https://doi.org/10.1007/s13399-022-02352-0

D. Mu, K. Ma, L. He, Z. Wei, Effect of microbial pretreatment on degradation of food waste and humus structure. Bioresour. Technol. **385**, 129442 (2023). https://doi.org/10.1016/j.biortech.2023.129442

S. Quintero-Herrera, P. Zwolinski, D. Evrard, J.J. Cano-Gómez, P. Rivas-García, Turning food loss and waste into animal feed: a Mexican spatial inventory of potential generation of agro-industrial wastes for livestock feed (2023). https://doi.org/10.1016/j.spc.2023.07.023

B.V. Raghunath, A. Punnagaiarasi, G. Rajarajan, A. Irshad, A. Elango, G. Mahesh Kumar, Impact of dairy effluent on environment—A review, in *Integrated Waste Management in India. Environmental Science and Engineering*, ed. by M. Prashanthi, R. Sundaram (Springer, Cham, 2016). https://doi.org/10.1007/978-3-319-27228-3_22

M.M. Ramirez-Corredores, M.R. Goldwasser, E. Falabella de Sousa Aguiar, Sustainable circularity, in *Decarbonization as a Route Towards Sustainable Circularity. SpringerBriefs in Applied Sciences and Technology* (Springer, Cham, 2023). https://doi.org/10.1007/978-3-031-19999-8_3

A.K. Rashwan, H. Bai, A.I. Osman, K.M. Eltohamym, Z. Chen, H.A. Younis, A. Al-Fatesh, D.W. Rooney, P.-S. Yap Recycling food and agriculture by-products to mitigate climate change: a review. Environ. Chem. Lett. (2023). https://doi.org/10.1007/s10311-023-01639-6

A. Rejeb, J.G. Keogh, K. Rejeb, Big data in the food supply chain: a literature review. J. of Data Inf. Manag. **4**, 33–47 (2022). https://doi.org/10.1007/s42488-021-00064-0

J.M. Rodríguez-Antón, L. Rubio-Andrada, M.S. Celemín-Pedroche, S.M. Ruíz-Peñalver, From the circular economy to the sustainable development goals in the European Union: an empirical comparison. Int. Environ. Agreements **22**, 67–95 (2022). https://doi.org/10.1007/s10784-021-09553-4

B. Sharma, R. Arora, G. Sharma, Alternative raw material selection and impact of goods and service tax in food industry. Mater. Today: Proc. **80**(1), 1–7 (2023). https://doi.org/10.1016/j.matpr.2022.09.221

J. Skinner, L. Shepstone, M. Hickson, A.A. Welch, Alcohol consumption and measures of sarcopenic muscle risk: cross-sectional and prospective associations within the UK biobank study. Calcif. Tissue Int. **113**, 143–156 (2023). https://doi.org/10.1007/s00223-023-01081-4

P. Staroń, Z. Kowalski, A. Staroń, M. Banach, Thermal treatment of waste from the meat industry in high scale rotary kiln. Int. J. Environ. Sci. Technol. **14**, 1157–1168 (2017). https://doi.org/10.1007/s13762-016-1223-9

J.F. Tahmassebi, A. BaniHani, Impact of soft drinks to health and economy: a critical review. Eur. Arch. Paediatr. Dent. **21**, 109–117 (2020). https://doi.org/10.1007/s40368-019-00458-0

N. Van Hung, M.C. Maguyon-Detras, M.V. Migo, R. Quilloy, C. Balingbing, P. Chivenge, M. Gummert, Rice straw overview: availability, properties, and management practices, in *Sustainable Rice Straw Management*, ed. by M. Gummert, N. Hung, P. Chivenge, B. Douthwaite (Springer, Cham, 2020). https://doi.org/10.1007/978-3-030-32373-8_1

A.H. Varnam, J.P. Sutherland, Soft drinks, in *Beverages* (Springer, Boston, MA, 1994). https://doi.org/10.1007/978-1-4615-2508-0_3

A. Veidemane, Education for sustainable development in higher education rankings: challenges and opportunities for developing internationally comparable indicators. Sustainability **14**, 5102 (2022). https://doi.org/10.3390/su14095102

A.K. Yadav, M.E. Khan, A. Pal, Biodiesel production from oleander (*Thevetia Peruviana*) oil and its performance testing on a diesel engine. Korean J. Chem. Eng. **34**, 340–345 (2017). https://doi.org/10.1007/s11814-016-0270-8

Z. Zhu, M. Gavahian, F.J. Barba, E. Roselló-Soto, D. Bursać Kovačević P. Putnik, G.I. Denoya, Valorization of waste and by-products from food industries through the use of innovative technologies, in *Agri-Food Industry Strategies for Healthy Diets and Sustainability*, ed. by F.J. Barba, P. Putnik, D. Bursać Kovačević (Elsevier, 2020). https://doi.org/10.1016/B978-0-12-817226-1.00011-4

Chapter 2
Anaerobic Digestion of Waste and By-Product from Food Industry

2.1 Introduction

The systematic development of civilisation is directly linked to the increasing demand for energy (Salehabadi et al. 2020). Consequently, there is a need to continuously increase its production in the form of electricity and heat. This energy will be used to meet the needs of the global economy and the municipal sector. The main source of energy production is the use of fossil fuels such as coal, oil and natural gas. However, it should be mentioned that their extraction and processing often burdens the environment and contributes to the greenhouse effect (Osman et al. 2023).

Another aspect to be highlighted is the limited fossil fuel resources. For this reason, other energy sources, including nuclear energy, are gaining importance (Danish and Pinter 2022). Among the advantages, the possibility of producing significant amounts of energy and the low environmental impact per unit of energy produced are mentioned first and foremost. Much controversy still exists on the subject of safety and the management of waste from nuclear energy production.

Renewable energy sources are a solution that is growing in importance every year (Miłek et al. 2022). Renewable energy, indicated as clean energy, comes from natural sources or processes that are constantly replenished. Wind energy, water energy and solar energy can be given as an example of renewable energy sources. However, the use of biomass plays a special role within renewable energy production (Das et al. 2020). Due to its high availability, biomass is the main source of human energy production in many places on Earth. This includes people in developing countries. Processing biomass can produce solid, liquid and gaseous biofuels. Within solid biofuels, briquettes and pellets are most commonly mentioned, and the raw materials for their production are mainly wood waste and straw (Czekała 2021). In the group of liquid biofuels, biodiesel or bioethanol can be mentioned. However, gaseous biofuels, including biogas, play a special role. They can be produced from a wide variety of substrates, including waste. Therefore, the type of biofuels in question has a special role in the context of waste and by-product management from food industry (Czekała 2018).

2.2 Anaerobic Digestion Process

Biodegradable waste has a special role in the overall waste sector. As the name suggests, these waste are biodegradable, i.e. decomposed under the influence of microorganisms, both under aerobic and anaerobic conditions—anaerobic digestion (Czekała et al. 2017). It is this characteristic that will be crucial in the context of their management and potential for environmental impact. Recently, particular attention has been given to the anaerobic digestion process as a waste management method. This process can produce both energy and a digestate that is an excellent fertiliser (Czekała 2022a, b). The number of substrates that can be used in biogas plants is considerable. It includes dozens of different groups of raw materials from which methane can be produced. For economic and logistical reasons, new raw materials are constantly being sought. In addition to the selection of raw materials, innovations and changes in the operation of biogas plants are also linked to the use of technology. Whatever measures are taken, it is important to bear in mind not only the cost-effectiveness of the solutions, but also the social and environmental aspects (Seberini 2020). All measures taken must also be implemented in accordance with the law.

The basis for the proper operation of a biogas plant is the selection of suitable substrates for energy production and their obtaining (Zamri et al. 2021). Substrates for the biogas plant should be delivered systematically every day (Czekała 2022a, b). Such an action will allow the biogas plant to function properly, also taking into account high efficiency. When the amount of biogas produced is high, this will translate into significant revenue from the sale of energy. In many parts of the world, the substrates used for biogas production have come from dedicated crops. One of the crops used for this purpose is maize. Maize silage is often a substrate for biogas production (Mazurkiewicz et al. 2019). Due to the conflict over the use of land for non-food purposes, in this case for energy production, this solution is slowly becoming less and less popular. Another factor in favour of this solution is the steadily increasing price of maize silage (2023a). Poland is a country where silage has been systematically losing its importance as a substrate for biogas production for several years now (Czekała et al. 2023b). The simplest solution to replace this substrate is to use readily available waste that can be used in facilities with the capacity to process it. These include unprocessed plant-based foodstuffs such as fruit and vegetables that have gone off the market, waste from markets and crop residues from households.

Increasingly, new substrates are being sought to produce as much biogas and methane as possible, while optimising the costs associated with their acquisition (Yoshida and Shimizu 2020). In this case, waste and by-product from food industry become the ideal solution. The raw materials in question are produced almost everywhere on Earth, which is directly linked to the demand for food (Czekała et al. 2016a). Another advantage of them in terms of anaerobic digestion is their diversity. This ensures that a variety of nutrients, including macronutrients and micronutrients, are delivered to the digester (Zielinski et al. 2019). This will provide the anaerobic digestion bacteria with the right conditions to multiply and produce biogas.

For many investors, the balance of substrate obtaining is a key advantage. Obtaining crops involves costs directly, most often an amount of several tens of euros or more. In the case of waste processing in biogas plants, there is no need to pay the costs of waste and by-products from food industry as for substrates from target crops, this can translate into additional revenue associated with their management. For the most problematic waste, e.g. that associated with the meat industry, it is possible to receive up to several hundred euros per tonne. Such a significant difference in substrate acquisition directly affects the profitability of agricultural biogas plants (Mazurkiewicz 2022). In this case, the use of waste and by-products from food industry, including problematic ones, for renewable energy production seems to be a reasonable and cost-effective solution. Combining the energy production sector with waste management will be a win–win solution for economic, environmental and social reasons.

Anaerobic digestion is a complex process occurring under the influence of microorganisms under anaerobic conditions. It is the absence of oxygen that is the main difference from composting (Bojarski et al. 2023). A characteristic aspect of anaerobic digestion is the participation of diverse groups of microorganisms, conditioning the biotechnological transformations that occur in four successive phases (Guo et al. 2014).

Anaerobic digestion is a complex process in which four stages can be identified:

(1) Hydrolysis—during this stage, sugars, proteins and fats are broken down to simpler compounds.
(2) Acidogenesis—during this stage, the products produced in the hydrolysis stage are broken down into fatty acids. This group primarily includes acetic acid, propionic acid, formic acid and butyric acid.
(3) Octanogenesis—during this stage, the organic acids produced in the previous stage are converted to acetic acid, carbon dioxide and hydrogen.
(4) Methanogenesis—during this stage, methane is produced (Janesch et al. 2021).

In order for the anaerobic digestion process to be effective, it is necessary to provide microorganisms with the necessary conditions for their functioning, growth and development (Nguyen et al. 2019). The most important activity is the selection of appropriate substrates that will allow for a large efficiency of methane. This group mainly includes feedstock rich in fats (Long et al. 2012), and waste from the meat industry is characterised by these properties. However, greater production of biogas and methane is associated with a longer time needed for decomposition. A different situation occurs in substrates rich in sugars (e.g. fruits and vegetables). Biogas production is lower, but decomposition time is much faster than in the case of substrates rich in proteins and fats. Individual substrates are characterised by properties that give some idea of their suitability as biogas feedstock. For the substrates used in the fermentation process, these will be primarily chemical composition, dry matter, organic matter content, biogas and methane efficiency and methane concentration in biogas. Selected parameters of the substrates related to the anaerobic digestion process along with a brief description are presented in Table 2.1.

Anaerobic digestion is a complex biological process. In order for methane production to be as high as possible and for the process not to stop, selected process parameters need to be systematically analysed and optimised (Madsen et al. 2011). The most important parameters of the anaerobic digestion process are: lack of oxygen, process temperature, type and intensity of mixing, dry matter, C/N ratio, pH, FOS/TAC and hydraulic retention time. In addition to these factors, the absence of oxygen and the content of process inhibitors should be monitored. Selected parameters related to the anaerobic digestion process are shown in Table 2.2.

The difference between composting and anaerobic digestion will be the products of the process. In the case of composting, the main product is compost, which is a fertiliser rich in organic matter (Czekała et al. 2016b). In the case of anaerobic digestion, the products will be biogas and digestate. Biogas is a mixture of gases formed as the most important product of the anaerobic decomposition of organic matter. The mixture consists mainly of methane, which is a source of energy (Alonso et al. 2020). Its concentration in biogas is usually above 50%, even reaching around 70%. For example, Ware and Power (2016) indicated a methane concentration of 69% for soft offal. The second main component is carbon dioxide, whose concentration can reach up to 50% and can be further valorised (Cordova et al. 2022). The

Table 2.1 Selected parameters of substrates used in anaerobic digestion

Properties of substrates	Description
Chemical composition	The chemical composition of the substrates used for biogas production is the most important aspect influencing their use in anaerobic digestion. It also influences the production of biogas, including methane. Feedstock for the biogas plant is a source of both organic matter and the macronutrients and micronutrients necessary for organisms to grow and develop
Dry matter	The dry matter content indicates the percentage of solids in the substrate analysed. The higher the value, the drier the substrate. The unit of dry matter content is %
Organic matter content	Organic matter content means the amount of solids in the residue lost during its combustion at 550 °C
Biogas efficiency	Biogas efficiency refers to the amount of biogas produced from 1 Mg of a substrate. The unit of biogas efficiency is $m^3\ Mg^{-1}$
Methane efficiency	Methane efficiency refers to the amount of methane produced from 1 Mg of a substrate. The unit of methane efficiency is $m^3\ Mg^{-1}$
Methane content	Methane concentration is the percentage of methane in the biogas produced

Table 2.2 Selected parameters related to the anaerobic digestion process

Anaerobic digestion process parameters	Description
Lack of oxygen	Anaerobic digestion is an anaerobic process. The methanogenic bacteria are absolute methanogens. The process only takes place properly in fully sealed digesters
Feedstock dry matter	The dry matter content of the digester can vary depending on the technology used. The most common assumption for wet fermentation is a maximum of 15% dry matter. For dry fermentation, the value of this parameter will be higher, at more than 30%
Temperature	Temperature is an important parameter affecting the anaerobic digestion process and the rate of substrate biodegradation. Most commonly, anaerobic digestion in terms of temperature is divided into psychrophilic, mesophilic and thermophilic. Regardless of the temperature at which the process is carried out, it should be constant. This applies particularly to mesophilic and thermophilic conditions
Mixing	The contents in the digester are not homogeneous. Mixing the digester contents prevents stratification. Systematic mixing allows for higher biogas production
C/N ratio	Each of the individual substrates has characteristic elemental contents. In order to provide microorganisms appropriate conditions, a proper C/N ratio must be ensured. Mozhiarasi and Natarajan (2022) state that the optimal C/N ratio is 25–30. A correct value for the C/N parameter will also reduce ammonia emissions
pH	Bacteria need the proper conditions to function. The pH is one of the most important parameters affecting the anaerobic digestion process. The value of this parameter is influenced by all the individual substrates that make up the biogas feedstock. Methanogenic bacteria prefer a pH of 6.8 to 7.5. Sometimes a separate hydrolysis phase, in which the pH is lower, is isolated in installations
FOS/TAC	It is the ratio of the intermediate alkalinity over the partial alkalinity. The first indicates the accumulation of volatile fatty acids, and second parameter indicates the alkalinity of the bicarbonates and is a measure of the buffer capacity in the digester. In English, it is also called IA/PA ratio
Hydraulic retention time	Hydraulic retention time is a parameter that describes how long substrates stay in the digester. Depending on the susceptibility to decomposition, the magnitude of this parameter can vary widely between substrates

production of significant amounts of CO_2 is characteristic of biomass degradation in both thermal and biological processes. In addition to these gases, ammonia and hydrogen sulphide can be mentioned most frequently. Nitrogen and hydrogen may also be formed in smaller quantities. These gases, although present in very small concentrations given in ppm, can affect the fermentation process and the combustion of biogas in the cogeneration engine. One way to use biogas is to combustion it in a cogeneration engine (Zhou et al. 2023). This will result in electricity and heat.

Purified and upgraded biogas can be injected in natural gas pipelines (Tomita et al. 2022).

Special attention should be paid to the second product of the anaerobic digestion process, which is the digestate. Digestate, otherwise known as digested pulp, is the residue not decomposed in the anaerobic digestion process (Czekała 2019). It mainly consists of water, undecomposed organic compounds, mineral compounds and biomass of organisms (Czekała et al. 2022). Depending on the legislation, digestate can be classified as a by-product, waste or fertiliser. Regardless of the legal classification, the substance in question is characterised by some specific properties that allow it to be used in several main ways. A popular option is to use the digestate as a fertiliser (Lohosha et al. 2023). Horta and Carneiro (2022) conducted a study to evaluate the fertilising value of digestate solid fraction as an organic amendment and as a source of nitrogen. The results indicated a positive effect on selected soil characteristics. Of particular interest is the separation of the digestate. This process results in two fractions with significantly different properties (Tambone et al. 2017). The liquid fraction is most often used as a fertiliser. The solid fraction, on the other hand, can be used as a fertiliser to increase the organic matter content of the soil, as well as for compost production (Czekała et al. 2023c). A particular direction for the use of the solid fraction of the digestate is its use for energy production (Czekała 2021).

2.3 Waste and By-Product from Food Industry as Substrates for Anaerobic Digestion Process

Waste and by-product from food industry, like food waste, are often used as animal feed. This solution has been in use for many years, although regulations in this area are becoming more stringent in many countries. According to van Raamsdonk et al. (2023), using former food products as feed ingredients is an important development in reducing bio-waste volumes and to achieve a footprint as small as possible for animal husbandry. Another management direction for this group of waste is direct use for fertiliser. It is also illegal, but practised to store substrates without the appropriate permits or to dispose of them in wrong places such as rivers or forests. Increasingly, however, waste of all kinds is being used in the energy production process in accordance with the waste-to-energy principle (Pluskal et al. 2022).

A distinction can be made between thermal and biological waste treatment methods. The high content of organic matter is an important aspect in both of these waste treatment groups. However, the excessive water content of most waste and by-products from food industry is the reason for treating these wastes using biological processes. A distinction can be made between anaerobic digestion and composting. The preferred direction is determined primarily by the properties of the waste and the financial possibilities of the investor. However, it should be mentioned that the properties of waste and by-products from food industry allow them to be used in both processes. Composting is a cheaper process (Alves et al. 2023), but the possibility of producing biogas and, further, energy speaks in favour of processing the raw materials in question in installations called biogas plants (Estévez et al. 2023).

In Polish law, agricultural biogas is defined as "gas obtained by anaerobic digestion of agricultural raw materials, agricultural by-products, liquid or solid animal excreta, by-products, waste or residues from the processing of products of agricultural origin or forest biomass, or plant biomass collected from areas other than those registered as agricultural or forest, excluding biogas obtained from raw materials originating from sewage treatment plants and landfills" (Act of February 20, 2015 on renewable energy sources, 2015). From the definition given above, it can be seen that numerous substrates can be used to produce agricultural biogas.

Feedstock for biogas plants is most often divided into substrates intentionally used for biogas production and waste. Within the first group, plants operating on raw materials of agricultural origin such as maize silage, straw or energy crops are mainly mentioned (Frankowski and Czekała 2023). In this case, the acquisition of substrates represents a cost. However, these substrates have a high potential for biogas production. Thus, the profit from the sale of energy and the additional benefits associated with renewable energy production (e.g. subsidies) compensate for the costs incurred. Natural fertilisers such as manure or slurry also complement the above-mentioned substrates (Tamburini et al. 2020). In particular, slurry has an important role in wet fermentation due to the fact of hydration. It should also be mentioned that raw materials from purpose-grown crops tend to have a longer decomposition time, with the associated increase in digester volume. Biogas plants are installations built for dedicated feedstock. For example, a sewage treatment plant will primarily use sewage sludge. In connection with municipal waste management, facilities for the anaerobic digestion of bio-waste or the fraction containing organic matter separated from mixed municipal waste will be useful.

The selection of substrates for any biogas plant is a key issue. In specific cases, the purchase of feedstock can account for up to 40% of the total operating costs of the plant. With a rational choice of feedstock for a biogas plant, the indicators in question can be positively influenced (Piwowar 2019). Furthermore, the costs associated with obtaining, e.g. maize silage can be replaced by revenues for accepting waste, e.g. that from the meat industry. The popularity of waste as substrates for biogas and biomethane production is growing every year. This applies in particular to waste from the agri-food industry. This is supported primarily by the price of obtaining them, which is usually a relatively low cost per Mg. Often, it is the owner of the waste who has to pay for its management. In this case, the prices offered by biogas plants are very favourable for waste producers. On the other hand, the positive effect of the waste in question on the anaerobic decomposition process should be emphasised. Due to the large variety of waste and by-products from food industry feedstock, a variety of nutrients will be supplied to the process. This limits the possibility of negative effects on the process resulting from deficiencies of certain elements. As reported by Demirel and Scherer, (2011), the availability of trace metals plays a significant role on the performances of agricultural biogas digesters operated with biomass. Among the elements playing a special role, the authors mention Fe, Co, Ni, Zn, Mo, W and Se.

Another extremely important issue concerning the operation of any biogas plant is the profitability of using specific substrates for energy production. Biodegradable waste plays a special role in this list. The most commonly used method is to compare the revenue from the sale of fuel or energy against the cost of obtaining the raw material. Taking into account a possible waste acceptance fee, the revenue per Mg of feedstock used can be estimated. This measure is the basis for calculating the balance sheet of the biogas plant. This type of estimation can be done in specialised biogas laboratories. The simplest method will be to carry out batch culture tests. These tests will result in information on biogas and methane production from 1 Mg of analysed feedstock and the methane concentration in the biogas. By comparing this with the cost/revenue resulting from processing the feedstock, indicators related to cost-effectiveness can be calculated. There are three key elements to consider when deciding on feedstock biogas. The first is related to their availability. In the case of integrated operation of a biogas plant with a feedstock plant, a significant proportion of the substrates is sourced locally. The second aspect relates to the sourcing of substrates from outside, which is related not only to the cost/revenue for sourcing them, but also to the cost of transport. This distance should not exceed 20–30 km. The last is directly related to the biogas efficiency of the feedstock and the concentration of methane in the biogas produced.

Activities related to the cost-effectiveness of installations are particularly important in the context of waste management. In this case, the costs associated with obtaining substrates are often not incurred. The situation can be different. Money can be raised for waste utilisation, which will have a positive effect on the balance of a biogas plant. Biogas plants are continuous plants. This means that biogas feedstock should be supplied to the plant every day (Metson et al. 2022). This action slowly achieves a stable biogas production and works with the assumed efficiency, which for a biogas plant can be even above 95%. This is done by properly selecting the substrates and delivering them to the biogas plant, which is a huge logistical challenge. For a medium-sized biogas plant delivering several tens of tonnes of substrate per day, it is important to organise the operations accordingly. An example of a biogas plant processing both manure and waste and by-products from food industry is shown in Fig. 2.1. In order to reduce the expenses associated with purchasing raw materials, it is recommended that waste and by-products from food industry are used for biogas production. Waste from this group is available everywhere as a result of the demand for food.

The number of substrates from the food industry is large. These substrates differ, among other things, in their physical and chemical properties and in their biogas and methane efficiency. Despite the fact that they come from completely different food production companies, almost all of them, if they contain organic matter, can be used. Regardless of this, it is advisable to keep substrate storage to a minimum. This is due to reasons such as the emissions that may occur during prolonged storage or warehousing. Another factor to be taken into account is that some wastes are produced throughout the year and others periodically. For example, beet pulp is a substrate that can only be obtained at a certain time. With these dependencies in mind, most often a biogas plant has contracted supplies from up to several dozen

Fig. 2.1 Biogas plant processing both manure and waste and by-products from food industry

suppliers. Exceptions can be biogas plants dedicated to specific plants, e.g. distilleries or dairies. In these cases, it may be the case that substrates are sourced only from a particular plant.

However, in order to reduce the costs associated with obtaining feedstock for biogas production and to improve the economic balance of biogas plants, alternative substrates for biogas production are being sought. One of the most important sources has turned out to be the agri-food industry, which has a large amount of waste, which in many biogas plants is the most commonly used substrate for energy production. In recent years, the number of installations working with the food sector has been increasing. They use numerous waste and by-products from food industry, as confirmed by numerous scientific studies. Selected studies are presented below.

Fruit and Vegetable Industry Waste and by-products from this sector represent one of the most diverse groups. This is influenced by both the wide variety of their types and the many types of waste generated in relation to them. The most common of the waste will be fruit and vegetables that have not made it to processing plants or have been rejected there, e.g. due to spoilage. According to Bancal and Ray (2022), 25–50% of fruits and vegetables are lost from farm to fork, together called post-harvest losses, on a global scale. A study by Czekała et al., (2016a) investigated the biogas and methane efficiency of selected fruits. Five fruits were analysed: apples, pears, plums, peaches and nectarines. Cumulative biogas production (fresh mass) ranged from 90.26 m^3 Mg^{-1} for peaches to 93.32 m^3 Mg^{-1} for pears. The methane concentration was also similar for the neutralised substrates and ranged from 51.80% for apples to 53.90% for pears. Other substrates in this group include fruit and vegetable residues, food processing waste, sludge from the processing of plant products, fruit and vegetable pomace, stones, peels, sludge, leaves, roots and stems, among others. Such a wide variety of waste and the location of processing plants in almost every part of the world offer the possibility to obtain them easily.

Grain Industry The production of biogas from cereal straw is one possibility for converting biomass into bioenergy. Due to its structure, straw may require pretreatment to increase biogas production and reduce retention time. In a study by Victorin et al. (2020), the biochemical methane production for untreated wheat straw was 237 NmL CH$_4$ g VS^{-1}. Other substrates classified in this group include grain, grain waste, husks and maize stalks. It should be noted that substrates from this group are often used for the production of solid biofuels such as briquettes. This is due to their high energy value and low water content. For example, the paper Alengebawy et al. (2023) states that the average values of approximate and ultimate analyses of rice straw in China are 10.78% moisture and 15.35 MJ kg^{-1} lower heating value.

Bakery and Confectionery Industry The waste in this group is characterised by a high dry matter content and considerable variation in the period when it begins to spoil. For example, dry waste can be stored for a relatively long time and creamy bakery waste for a short time. Within this group, waste such as stale bread, bread rolls, cookies, confectionery waste, creamy and non-creamy bakery waste, beet pulp and catering waste can be mentioned. Konkol et al. (2018) conducted research on selected waste from this group. Wheat roll, wheat bread, rye bread and bun were used in the study. Cumulative biogas in terms of volatile solids was 839.3 m^3 Mg^{-1} for wheat roll, 784.2 m^3 Mg^{-1} for wheat bread, 859.6 m^3 Mg^{-1} for rye bread and the highest 1283.5 m^3 Mg^{-1} for bun, respectively. Methane content ranged from 45.8% for wheat bread to 52.1% for bun.

Meat Industry Waste and by-products from meat industry constitute a very important group for biogas feedstock. A wide variety of substrates can be listed in this group, including slaughterhouse waste, gastric content, fats, fatty sludge, protein and fat waste, bones, blood, skin, fats, hair, bristles, feathers, intestines, horns and hooves. Among the greatest advantages of the substrates in this group are the very high potential for biogas and methane production. This is due to the significant amounts of fats and proteins in the waste. The biggest challenge, however, is the need for pretreatment of the raw materials, which is primarily aimed at reducing biological risks. This involves costs for additional equipment and increased energy consumption, but this is offset by the benefits of waste acceptance and high biogas production. As highlighted by Mozhiarasi and Natarajan (2022), the C/N ratio of slaughterhouse and poultry waste is found between 7 and 10. This may prove to be inhibitory for aaerobic digestion, where C/N is preferred at 25–30. Hence, the need arises to use additional substrates for digestion.

Fishing Industry Waste from fish processing plants mainly include heads, fish guts and wastewater. As with waste and by-product from meat industry, these are substrates that allow high biogas and methane efficiency. However, attention should be paid to the scale of production. In many countries, the fish industry is a relatively small branch of the food industry. Poland is a case in point. On the other hand, in countries like Japan, fish consumption is very large, which translates into a significant amount of waste. Ivanovs et al. (2018) report that research on the fish waste anaerobic digestion shows a biomethane production potential of 0.2–0.9 CH_4 m^3 kg^{-1} VS.

Dairy Industry Waste and by-products from dairy industry are characterised by a relatively low dry matter content. For this reason, appropriate methods must be adopted to manage them. Substrates in this group include whey, permeate, buttermilk and residues from milk, cottage cheese and cheese production and wastewater. The substrates in question often have high chemical oxygen demand values. Anaerobic technology is interesting because it allows a significant reduction in this parameter. In the dairy industry, significant amounts of water are used, which involves a large amount of wastewater. Therefore, the wastewater generated from this group is treated. Kozlowski et al. (2019) analysed three significantly different waste and by-products from dairy industry due to their biogas potential. The dry matter content for whey was 6.38%, for dairy sludge 12.42% and for fat sludge was as high as 45.67%. As well as dry matter, the organic matter content for the individual substrates varied. The lowest content of 65.86% was for fat sludge and the highest for whey (92.32%). Dairy sludge was characterised by an organic matter content of 82.86%. Cumulated biogas production for whey was 45.53 m^3 Mg^{-1} FM and for dairy sludge 49.04 m^3 Mg^{-1}. Fat sludge was characterised by a significantly higher biogas efficiency of as much as 349.61 m^3 Mg^{-1}. This is due to the fact that fatty waste is characterised by a higher production of not only biogas, but also methane, as mentioned, e.g. in the section on waste and by-products from meat industry. This was also confirmed by the high methane concentration for fat sludge 61.89%, against 52.78% for whey and 57.43% for dairy sludge.

Oil and Fat Industry This group includes waste and by-products such as waste from the production of vegetable oil, post-extraction meal and sludge. Oil cakes are by-products of oil processing and are rich in nutrients. The article by Mohanty et al. (2022) presented the composition of major oil cakes, including: soya bean cakes, reapeseed cakes, cottonseed cakes, groundnut cakes, sunflower seed cakes, palm kernel cakes and copra cakes. The dry matter of the raw materials listed ranged from 84.8% (soybean cakes) to 94.3% (cottonseed cakes). It should be mentioned that more and more oil cakes involve non-edible plants capable of producing oils. The authors quoted above-mentioned plants such as Jatropha (*Jatropha curcas*), karanja (*Pongamia pinnata*), mahua (*Madhuca indica*), silk cotton tree (*Ceiba pentandra*) and castor (*Ricinus communis*) are the non-edible plants capable of producing oils. This is related, among other things, to the production of biodiesel.

Soft Drink Industry Waste and by-products from soft drink industry will be very similar to those substrates generated for the fruit and vegetable industry. These will therefore include fruit and vegetables unfit for consumption and fruit and vegetable pomace. Other substances are produced, for example, in the production of non-alcoholic beer, but this will be discussed in the next group. Waste and by-products from soft drink industry are a source of many compounds, among which are carbohydrates, proteins, mineral compounds and vitamins. These substrates are often used as animal feed, but both composting and anaerobic digestion are also popular management options. Neto et al. (2021) researched cumulative biogas production for food, fruit and vegetable waste. The database was built on mixed values of eight variables seen in literature. Research on the topic in question was conducted by Czubaszek et al. (2022). The study aims to determine the specific methane efficiency of anaerobic digestion of four fruit and vegetable residues: apple pomace, cabbage leaves, pumpkin peels and fibrous strands and walnut husks. Methane production for apple pomace was 232.20 NL kg VS^{-1}.

Alcoholic Beverage Industry The generation of waste classified in this group is primarily related to the production of beer, wine and spirits. Brewer's malt is associated with beer production, pomace and sludge with wine production and distillers grains with spirits. Sganzerla et al. (2023) conducted research on the anaerobic co-digestion of brewery by-products for biomethane and bioenergy recovery. The biochemical methane potential was conducted under mesophilic conditions by mixing brewery wastewater and sludge from the brewery wastewater treatment plant (1:1, v/v), following the addition of brewer's spent grains in the range of 2.5–12.5%, w/v. The study showed that the addition of this substrate increased methane efficiency. As the authors point out, the data obtained provide a perspective on using brewery by-products for bioenergy production, as a part of circular bioeconomy transition of the beer industry.

Feed Industry Feeds are produced from a variety of substrates that can be divided into plant and animal origin. Waste and by-products from this group include mainly unused parts of raw materials. The methane efficiency of the substrates in this group will vary greatly, which is mainly due to the type of raw materials used for biogas production.

   On the basis of the information presented above on the ten groups discussed, it should be noted that the diversity of biogas feedstock is large. The use of waste and by-products from the food industry is gaining in importance. This is due, among other things, to the desire to recycle waste rather than landfill and the possibility of gaining revenue for accepting waste, especially problematic waste such as from the meat industry. Changes in the trend of using raw materials for biogas production can be illustrated using the example of agricultural biogas plants in Poland. For this purpose, data on substrates used for agricultural biogas production in 2011 (Table 2.3) and 2022 (Table 2.4) were compared.

   In 2011, 469,416 Mg of substrates was used for agricultural biogas production in Poland. Slurry accounted for less than 57% of these raw materials by weight. Maize silage was ranked second, with a share of just over 23%. The data presented show that these two most popular feedstocks at that time accounted for approximately 80%

Table 2.3 List of substrates used for agricultural biogas production in 2011, compiled based on quarterly reports submitted to the National Support Centre for Agriculture (KOWR) (2011)

No.	Substrates	Amount (Mg)
1.	Slurry	265,960.790
2.	Maize silage	108,876.140
3.	Distillers grains	30,465.110
4.	Manure	11,640.530
5.	Fruit and vegetable residues	10,984.350
6.	A mixture of lecithin and soaps	8906.870
7.	Potato pulp	7258.490
8.	Grass silage	7217.100
9.	Beet pulp	6922.450
10.	Cereal silage	5973.800
11.	Waste from the dairy industry	1933.000
12.	Cereal	1611.770
13.	Gastric contents	1278.300
14.	Fatty waste	285.650
15.	Flour, breadcrumbs and batter	101.710
	Total	**469,416.060**

of the feedstock for Polish agricultural biogas plants (Table 2.3). The share of other feedstocks, including waste and by-products, from food industry was less significant.

This situation has systematically changed over the years. This was related not only to increased awareness, but also to unfavourable conditions on the domestic market regarding the support system for renewable energy. Biogas plant owners, in an attempt to increase revenue, started accepting more and more waste at the plants. At the same time, the amount of maize silage, which was expensive to purchase, was reduced. As a result, over the next few years, the share of all kinds of waste and by-products from food industry used for agricultural biogas production increased steadily. This is confirmed by the data presented in Tables 2.3 and 2.4. In 2022, 5,696,273 Mg of substrates was used for agricultural biogas production in Poland, and this was more than 10 times more than in 2011. This confirms the development of the domestic biogas market. Slurry and maize silage, which ranked first and second in 2011, will take second and fifth place, respectively, in 2022. Distillers grains ranked first (1,068,753 Mg), accounting for less than 19% of the feedstock used in 2022. Many waste and by-products such as food processing waste (3rd place) and fruit and vegetable residues (4th place) appeared in the list. The list of substrates used for agricultural biogas production in 2022, compiled based on quarterly reports submitted to the National Support Centre for Agriculture (KOWR), is presented in Table 2.4.

Some of the raw materials obtained will require pretreatment (Vieira et al. 2022). Pretreatment allows prepared substrate for the process. It may be required, for example, due to current legislation, e.g. thermal treatment of waste from the meat

Table 2.4 List of substrates used for agricultural biogas production in 2022, compiled based on quarterly reports submitted to the National Support Centre for Agriculture (KOWR) (2022)

No.	Substrates	Amount (Mg)
1.	Distillers grains	1,068,753.900
2.	Slurry	934,770.175
3.	Food processing waste	781,626.329
4.	Fruit and vegetable residues	773,408.075
5.	Maize silage	612,970.973
6.	Technological sludge from the agri-food industry	308,682.286
7.	Beet pulp	229,875.868
8.	Waste from the dairy industry	173,703.642
9.	Expired food	161,101.120
10.	Slaughterhouse waste	125,158.887
11.	Manure	97,262.251
12.	Waste plant	75,684.360
13.	Fruits and vegetables	48,545.176
14.	Cereal, cereal waste	46,504.850
15.	Poultry manure	46,435.205
16.	Fats	39,332.375
17.	Sludge from the processing of plant products	36,850.558
18.	Grass and cereal silage	34,170.563
19.	Green fodder	18,750.243
20.	Slops	15,924.520
21.	Gastric contents	14,003.320
22.	Waste from the production of vegetable oil	9548.880
23.	Fodder	8960.800
24.	Straw	8005.809
25.	Extraction pomace from the production of herbal pharmaceuticals	7118.680
26.	Vegetable oils	6099.710
27.	Protein and fat waste	5268.400
28.	Digestate	3119.290
29.	Fatty sludge	1864.580
30.	Liquid wheat residues	1145.363
31.	Coffee	802.810
32.	Catering waste	658.506
33.	Glycerine	92.920
34.	A mixture of lecithin and soaps	48.740
35.	Protein and fat slurries	24.460
	Total	**5,696,273.624**

industry. Another example is the grinding of substrates to obtain a higher biogas production due to easier access of bacteria to organic matter. The most common groups of pretreatment methods are as follows:

(1) physical, e.g. mechanical, thermal,
(2) chemical, e.g. with acids or alkalis,
(3) biological, e.g. microbiological or using enzymes,
(4) mixed methods, e.g. thermal-chemical.

Regardless of the chosen method, the main aim of pretreatment is to increase the biogas and methane efficiency of the substrates. However, care must be taken to ensure that the energy production from biogas is higher than the energy input and costs associated with the treatment process itself. Sometimes, the use of specific solutions does not produce the expected results. For this reason, some research are being carried out to find the most optimal method. Nowicka et al. (2021) found that the substrate heating method affected the volume of biogas produced. In a study by Kovács et al. (2022), lignocellulosic substrates were subjected to pretreatment by hydrolytic enzymes. As reported by the authors substantially increased biogas production was obtained from the pretreated substrates, with most efficient strain for pretreatment and biogas using *Penicillium aurantiogriseum*. If these are not possible or not cost-effective, other measures are recommended to positively influence the financial balance of the biogas plant. The most commonly practised solution is the selection of waste, which allows an adequate efficiency of biogas while charging for its management.

Waste and by-products from food industry generally have a stable composition that is well and predictable from one batch of waste to the next. This is very important because of the possibility of their use not only in biogas plants operating at the plants producing them. Some of these raw materials can be used in typical agricultural biogas plants, e.g. together with slurry and silage, without changing the status of the plant to waste. This is extremely important and practical, as it allows owners of agricultural biogas plants to source new feedstocks and waste producers to recycle them without having their own biogas plant. When analysing aspects related to feedstock obtaining and the optimisation of the anaerobic digestion process, biogas use should be highlighted. Biogas can be used in a number of ways, allowing the optimal solution to be selected for each plant (Mertins and Wawer 2022).

2.4 Conclusion

Anaerobic digestion process is one of the ways to produce renewable energy. Furthermore, not only targeted crops, but also waste such as waste and by-products from food industry can be used for biogas production. Food production facilities are increasingly investing in biogas plants, which are suitable for on-site management of biodegradable waste. In this case, there is no charge for waste management to another company. Another advantage is a profit from selling energy or reducing the cost of energy from suppliers. Biogas can be used, among other things, to produce electricity and heat in cogeneration or to biomethane production. It should be noted that biogas plants are stable energy sources, independent of weather conditions. Among the most important challenges for their operation are the regular supply of biogas feedstock and the care of the anaerobic digestion process taking place in the digesters. By implementing these two factors at a high level, the efficiency of a biogas plant can be close to 100%, making a positive contribution to the energy management of plants in the agri-food industry. This is one of the biggest advantages over unstable energy sources such as wind and solar power.

References

Act of February 20, 2015 on renewable energy sources (consolidated text). https://isap.sejm.gov.pl/isap.nsf/DocDetails.xsp?id=WDU20150000478 Accessed 31 July 2023 (in Polish)

A. Alengebawy, Y. Ran, N. Ghimire, A.I. Osman, P. Ai, Rice straw for energy and value-added products in China: a review. Environ. Chem. Lett. (2023). https://doi.org/10.1007/s10311-023-01612-3

M.J.C. Alonso, M.R.M. Franco, A.P. Valenciana, L.E. Montañez-Hernández, Metabolic engineering: a tool to increase the methane yield and efficiency of anaerobic digestion process, in *Biogas Production*, ed. by N. Balagurusamy, A.K. Chandel (Springer, Cham, 2020). https://doi.org/10.1007/978-3-030-58827-4_11

D. Alves, I. Villar, S. Mato, Community composting strategies for biowaste treatment: methodology, bulking agent and compost quality. Environ. Sci. Pollut. Res. (2023). https://doi.org/10.1007/s11356-023-25564-x

V. Bancal, R.C. Ray, Overview of food loss and waste in fruits and vegetables: from issue to resources, in *Fruits and Vegetable Wastes*, ed. by R.C. Ray (Springer, Singapore, 2022). https://doi.org/10.1007/978-981-16-9527-8_1

W. Bojarski, W. Czekała, M. Nowak, J. Dach, Production of compost from logging residues. Bioresour. Technol. **376**, 128878 (2023). https://doi.org/10.1016/j.biortech.2023.128878

S.S. Cordova, M. Gustafsson, M. Eklund, N. Svensson, Potential for the valorization of carbon dioxide from biogas production in Sweden. J. Clean. Prod. **370**, 133498 (2022). https://doi.org/10.1016/j.jclepro.2022.133498

W. Czekała, Solid fraction of digestate from biogas plant as a material for pellets production. Energies **14**(16), 5034 (2021). https://doi.org/10.3390/en14165034

W. Czekała, Digestate as a source of nutrients: nitrogen and its fractions. Water **14**(24), 4067 (2022a). https://doi.org/10.3390/w14244067

W. Czekała, T. Jasiński, M. Grzelak, K. Witaszek, J. Dach, Biogas plant operation: digestate as the valuable product. Energies **15**, 8275 (2022b). https://doi.org/10.3390/en15218275

W. Czekała, T. Jasiński, J. Dach, Profitability of the operation of agricultural biogas plants in Poland, depending on the substrate use model. Energy Rep. **9**, 196–203 (2023a). https://doi.org/10.1016/j.egyr.2023.05.175

W. Czekała, J. Pulka, T. Jasiński, P. Szewczyk, W. Bojarski, J. Jasiński, Waste as substrates for agricultural biogas plants: A case study from Poland. J. Water Land Dev. **56**(I–III):1–6 (2023b). https://doi.org/10.24425/jwld.2023.143743

W. Czekała, M. Nowak, G. Piechota, Sustainable management and recycling of anaerobic digestate solid fraction by composting: a review. Bioresour. Technol. **375**, 128813 (2023c). https://doi.org/10.1016/j.biortech.2023.128813

W. Czekała, A. Smurzyńska, M. Cieślik, P. Boniecki, K. Kozłowski, Biogas efficiency of selected fresh fruit covered by the Russian embargo, in *Energy And Clean Technologies Conference Proceedings, SGEM 2016*, vol III (2016a), pp. 227–233. https://doi.org/10.5593/sgem20 16HB43

W. Czekała, J. Dach, D. Janczak, A. Smurzyńska, A. Kwiatkowska, K. Kozłowski, Influence of maize straw content with sewage sludge on composting process. J. Water Land Dev. **30**, 43–49 (2016b). https://doi.org/10.1515/jwld-2016-0020

W. Czekała, K. Gawrych, A. Smurzyńska, J. Mazurkiewicz, A. Pawlisiak, D. Chełkowiski, M. Brzoski, The possibility of functioning micro-scale biogas plant in selected farm. J. Water Land Dev. **35**(X–XII), 19–25 (2017). https://doi.org/10.1515/jwld-2017-0064

W. Czekała, agricultural biogas plants as a chance for the development of the agri-food sector. J. Ecol. Eng. **19**(2), 179–183 (2018). https://doi.org/10.12911/22998993/83563

W. Czekała, Biogas production from raw digestate and its fraction. J. Ecol. Eng. **20**(6), 97–102 (2019). https://doi.org/10.12911/22998993/108653

W. Czekała, Biogas as a sustainable and renewable energy source, in *Clean Fuels for Mobility. Energy, Environment, and Sustainability*, ed. by G. Di Blasio, A.K. Agarwal, G. Belgiorno, P.C. Shukla (Springer, Singapore, 2022). https://doi.org/10.1007/978-981-16-8747-1_10

R. Czubaszek, A. Wysocka-Czubaszek, R. Tyborowski, Methane production potential from apple pomace, cabbage leaves, pumpkin residue and walnut husks. Appl. Sci. **12**, 6128 (2022). https://doi.org/10.3390/app12126128

M.S.S. Danish, G. Pinter, Environmental and economic efficiency of nuclear projects, in *Clean Energy Investments for Zero Emission Projects. Contributions to Management Science*, ed. by H. Dinçer, S. Yüksel (Springer, Cham, 2022). https://doi.org/10.1007/978-3-031-12958-2_10

P.K. Das, B.P. Das, P. Dash, Role of energy crops to meet the rural energy needs: an overview, in *Biomass Valorization to Bioenergy. Energy, Environment, and Sustainability*, ed. by R. Praveen Kumar, B. Bharathiraja, R. Kataki, V. Moholkar (Springer, Singapore, 2020). https://doi.org/10.1007/978-981-15-0410-5_2

B. Demirel, P. Scherer, Trace element requirements of agricultural biogas digesters during biological conversion of renewable biomass to Methane. Biomass Bioenerg. **35**(3), 992–998 (2011). https://doi.org/10.1016/j.biombioe.2010.12.022

S. Estévez, R. Rebolledo-Leiva, D. Hernández, S. González-García, G. Feijoo, M.T. Moreira, Benchmarking composting, anaerobic digestion and dark fermentation for apple vinasse management as a strategy for sustainable energy production. Energy **274**, 127319 (2023). https://doi.org/10.1016/j.energy.2023.127319

J. Frankowski, W. Czekała, Agricultural plant residues as potential co-substrates for biogas production. Energies **16**(11), 4396 (2023). https://doi.org/10.3390/en16114396

X. Guo, C. Wang, F. Sun, W. Zhu, W. Wu, A comparison of microbial characteristics between the thermophilic and mesophilic anaerobic digesters exposed to elevated food waste loadings. Bioresour. Technol. **152**, 420–428 (2014). https://doi.org/10.1016/j.biortech.2013.11.012

C. Horta, J.P. Carneiro, Use of digestate as organic amendment and source of nitrogen to vegetable crops. Appl. Sci. **12**, 248 (2022). https://doi.org/10.3390/app12010248

K. Ivanovs, K. Spalvins, D. Blumberga, Approach for modelling anaerobic digestion processes of fish waste. Energy Procedia **147**, 390–396 (2018). https://doi.org/10.1016/j.egypro.2018.07.108

E. Janesch, J. Pereira, P. Neubauer, S. Junne, Phase separation in anaerobic digestion: a potential for easier process combination? Front. Chem. Eng. Sec. Environ. Chem. Eng. **3** (2021). https://doi.org/10.3389/fceng.2021.711971

I. Konkol, L. Świerczek, A. Cenian, Biogas production from bakery wastes—Dynamics, retention time and biogas potential. J. Agric. Eng. **63**(1) (2018)

E. Kovács, C. Szűcs, A. Farkas, M. Szuhaj, M. Maróti, Z. Bagi, G. Rákhely, K.L. Kovács, Pretreatment of lignocellulosic biogas substrates by filamentous fungi. J. Biotechnol. **360**(10), 160–170 (2022). https://doi.org/10.1016/j.jbiotec.2022.10.013

K. Kozłowski, M. Pietrzykowski, W. Czekała, J. Dach, A. Kowalczyk-Juśko, K. Jóźwiakowski, M. Brzoski, Energetic and economic analysis of biogas plant with using the dairy industry waste. Energy **183**, 1023–1031 (2019). https://doi.org/10.1016/j.energy.2019.06.179

National Support Centre for Agriculture (KOWR), List of substrates used for agricultural biogas production in 2011, compiled based on quarterly reports submitted to the National Support Centre for Agriculture (KOWR) (2011)

National Support Centre for Agriculture (KOWR), List of substrates used for agricultural biogas production in 2022, compiled based on quarterly reports submitted to the National Support Centre for Agriculture (KOWR) (2022)

R. Lohosha, V. Palamarchuk, V. Krychkovskyi, Economic efficiency of using digestate from biogas plants in Ukraine when growing agricultural crops as a way of achieving the goals of the European Green Deal. Energy Policy **26**(2), 161–182 (2023). https://doi.org/10.33223/epj/163434

J.H. Long, T.N. Aziz, F.L. de los Reyes III, J.J. Ducoste, Anaerobic co-digestion of fat, oil, and grease (FOG): a review of gas production and process limitations. Process Saf. Environ. Prot. **90**(3), 231–245 (2012). https://doi.org/10.1016/j.psep.2011.10.001

M. Madsen, J.B. Holm-Nielsen, K.H. Esbensen, Monitoring of anaerobic digestion processes: a review perspective. Renew. Sust. Energ. Rev. **15**(6), 3141–3155 (2011). https://doi.org/10.1016/j.rser.2011.04.026

J. Mazurkiewicz, Energy and economic balance between manure stored and used as a substrate for biogas production. Energies **15**, 413 (2022). https://doi.org/10.3390/en15020413

J. Mazurkiewicz, A. Marczuk, P. Pochwatka, S. Kujawa, Maize straw as a valuable energetic material for biogas plant feeding. Materials **12**, 3848 (2019). https://doi.org/10.3390/ma12233848

A. Mertins, T. Wawer, How to use biogas?: A systematic review of biogas utilization pathways and business models. Bioresour. Bioprocess. **9**, 59 (2022). https://doi.org/10.1186/s40643-022-00545-z

G.S. Metson, A. Sundblad, R. Feiz, N.-H. Quttineh, S. Mohr, Swedish food system transformations: rethinking biogas transport logistics to adapt to localized agriculture. Sustain. Prod. Consum. **29**, 370–386 (2022). https://doi.org/10.1016/j.spc.2021.10.019

D. Miłek, P. Nowak, J. Latosińska, The development of renewable energy sources in the European Union in the light of the European Green Deal. Energies **15**, 5576 (2022). https://doi.org/10.3390/en15155576

A. Mohanty, P.R. Rout, B. Dubey, S. Singh Meena, P. Pal, M. Goel, A critical review on biogas production from edible and non-edible oil cakes. Biomass Conv. Bioref. **12**, 949–966 (2022). https://doi.org/10.1007/s13399-021-01292-5

V. Mozhiarasi, T.S. Natarajan, Slaughterhouse and poultry wastes: management practices, feedstocks for renewable energy production, and recovery of value added products. Biomass Conv. Bioref. (2022). https://doi.org/10.1007/s13399-022-02352-0

J.G. Neto, L.V. Vidal Ozorio, T.C. Campos de Abreu, B.F. dos Santos, F. Pradelle, Modeling of biogas production from food, fruits and vegetables wastes using artificial neural network (ANN). Fuel **285**, 119081 (2021). https://doi.org/10.1016/j.fuel.2020.119081

L.N. Nguyen, A.Q. Nguyen, L.D. Nghiem, Microbial community in anaerobic digestion system: progression in microbial ecology, in *Water and Wastewater Treatment Technologies. Energy, Environment, and Sustainability*, ed. by X.T. Bui, C. Chiemchaisri, T. Fujioka, S. Varjani (Springer, Singapore, 2019). https://doi.org/10.1007/978-981-13-3259-3_15

A. Nowicka, M. Zieliński, M. Dębowski, M. Dudek, Progress in the production of biogas from maize silage after acid-heat pretreatment. Energies **14**, 8018 (2021). https://doi.org/10.3390/en14238018

A.I. Osman, L. Chen, M. Yang, G. Msigwa, M. Farghali, S. Fawzy, D.W. Rooney, P.S. Yap, Cost, environmental impact, and resilience of renewable energy under a changing climate: a review. Environ. Chem. Lett. **21**, 741–764 (2023). https://doi.org/10.1007/s10311-022-01532-8

A. Piwowar, Development of the agricultural biogas market in Poland—Production volume, feedstocks, activities and behaviours of farmers. Sci. J. Wars. Univ. Life Sci.–SGGW Probl. World Agric. **19**, 88–97 (2019). https://doi.org/10.22630/PRS.2019.19.1.8

J. Pluskal, R. Šomplák, D. Hrabec, V. Nevrlý, L.M. Hvattum, Optimal location and operation of waste-to-energy plants when future waste composition is uncertain. Oper Res Int J **22**, 5765–5790 (2022). https://doi.org/10.1007/s12351-022-00718-w

A. Salehabadi, M.I. Ahmad, N. Ismail, N. Morad, M. Enhessari, Overview of energy, society, and environment towards sustainable and development, in *Energy, Society and the Environment. SpringerBriefs in Applied Sciences and Technology* (Springer, Singapore, 2020). https://doi.org/10.1007/978-981-15-4906-9_1

A. Seberini, Economic, social and environmental world impacts of food waste on society and Zero waste as a global approach to their elimination, in *SHS Web of Conferences*, vol. 74 (2020), p. 03010. https://doi.org/10.1051/shsconf/20207403010

W.G. Sganzerla, M. Tena, L. Sillero, F.E. Magrini, I.V. Machado Sophiatti, J. Gaio, S. Paesi, T. Forster-Carneiro, R. Solera, M. Perez, Application of anaerobic co-digestion of brewery by-products for biomethane and bioenergy production in a biorefinery concept. Bioenerg. Res (2023). https://doi.org/10.1007/s12155-023-10605-7

F. Tambone, V. Orzi, G. D'Imporzano, F. Adani, Solid and liquid fractionation of digestate: mass balance, chemical characterization, and agronomic and environmental value. Bioresour. Technol. **243**, 1251–1256 (2017). https://doi.org/10.1016/j.biortech.2017.07.130

E. Tamburini, M. Gaglio, G. Castaldelli, E.A. Fano, Biogas from agri-food and agricultural waste can appreciate agro-ecosystem services: the case study of Emilia Romagna Region. Sustainability **12**, 8392 (2020). https://doi.org/10.3390/su12208392

E. Tomita N, Kawahara, U. Azimov, Significance of biogas, its production and utilization in gas engines, in *Biogas Combustion Engines for Green Energy Generation. SpringerBriefs in Applied Sciences and Technology* (Springer, Cham, 2022). https://doi.org/10.1007/978-3-030-94538-1_1

L.W.D. Van Raamsdonk, N. Meijer, E.W.J. Gerrits, M.J. Appel, New approaches for safe use of food by-products and biowaste in the feed production chain. J. Clean. Prod. **388**, 135954 (2023). https://doi.org/10.1016/j.jclepro.2023.135954

M. Victorin, A. Davidsson, O. Wallberg, Characterization of mechanically pretreated wheat straw for biogas production. Bioenerg. Res. **13**, 833–844 (2020). https://doi.org/10.1007/s12155-020-10126-7

S. Vieira, J. Schneider, W.J. Martinez Burgos, A. Magalhães A.B. Pedroni Medeiros, J.C. de Carvalho, L.P. de Souza Vandenberghe, C.R. Soccol, E.B. Sydney, Pretreatments of solid wastes for anaerobic digestion and its importance for the circular economy, in *Handbook of Solid Waste Management*, ed. by C. Baskar, S. Ramakrishna, S. Baskar, R. Sharma, A. Chinnappan, R. Sehrawat (Springer, Singapore, 2022). https://doi.org/10.1007/978-981-16-4230-2_5

A. Ware, N. Power, Biogas from cattle slaughterhouse waste: energy recovery towards an energy self-sufficient industry in Ireland. Renew. Energy **97**, 541–549 (2016). https://doi.org/10.1016/j.renene.2016.05.068

K. Yoshida, N. Shimizu, Biogas production management systems with model predictive control of anaerobic digestion processes. Bioprocess Biosyst. Eng. **43**, 2189–2200 (2020). https://doi.org/10.1007/s00449-020-02404-7

M.F.M.A. Zamri, S. Hasmady, A. Akhiar, F. Ideris, A.H. Shamsuddin, M. Mofijur, I.M. Rizwanul Fattah, T.M.I. Mahlia, A comprehensive review on anaerobic digestion of organic fraction of municipal solid waste. Renew. Sust. Energ. Rev. **137**, 110637 (2021). https://doi.org/10.1016/j.rser.2020.110637

J. Zhou, M. Ashraf Ali, A.M. Hussein Wais, S. Fahad Almojil A.I. Almohana, A. Fahmi Alali, M.R. Ali, M. Sohail, A novel modified biogas-driven electricity/cooling cogeneration system using open-and-closed Brayton cycle concepts: environmental analysis and optimization. Ain Shams Eng. J. 102230 (2023) (in press, corrected proof)

M. Zieliński, M. Kisielewska, M. Dębowski, K. Elbruda, Effects of nutrients supplementation on enhanced biogas production from maize silage and cattle slurry mixture. Water Air Soil Pollut. **230**, 117 (2019). https://doi.org/10.1007/s11270-019-4162-5

Chapter 3
Composting of Waste and By-Product from Food Industry

3.1 Introduction

Biodegradable waste can be produced in agriculture, municipal sector and industrial sector. Despite the many different sources of origin, their selected properties are similar. The most important of these is a high organic matter content and often, although not always, a high water content (Czekała et al. 2023). For example, substrates with a high moisture content can be slurry, distillery grains or wastewater from food industry. On the other hand, biomass with a relatively low water content that is biodegradable can be straw, maize stalks or brewery by-products. In addition to using the waste and by-products in question in the anaerobic digestion process, they can be recycled in the composting process (Sharma et al. 2023).

Composting is one of the biological waste treatment methods. As a result of the composting process, the substrates used are degraded, and the product will be composted—high-quality organic fertiliser. This process takes place under aerobic conditions with the participation of microorganisms. These microorganisms are different from those used in anaerobic digestion because; in contrast to decomposition without oxygen, oxygen is essential in composting (Zhou et al. 2020).

Waste and by-products from food industry are in many cases suitable substrates for compost production (Czekała et al. 2022a, b). However, the amount of compostable substrates from the group in question is less than for anaerobic digestion. The lack of possibility of use or limitation concerns, among others, waste from the meat sector. This is related to the possibility of spreading disease. Another example would be the lack of profitability of the use of some wastewater from on-site treatment plants. This is due to the low dry matter content. In this case, however, the solid fraction obtained from wastewater separation can be composted. This is similar to the separation of solids from liquid slurry. Despite some limitations, composting is a popular method for treating biodegradable waste (Mozhiarasi and Natarajan 2022).

W. Czekała, *Biological Treatment of Waste and By-Products from Food Industry*, SpringerBriefs in Applied Sciences and Technology, https://doi.org/10.1007/978-3-031-47487-3_3

3.2 Composting Process

Characteristics of the Process

Composting is one of the most common methods for treating biodegradable waste (Bojarski et al. 2023). Most waste and by-products from food industry provide a proper substrate for this process. The technology has been known and used for many years. Composting can be defined as an autothermic and thermophilic decomposition process of organic matter occurring under aerobic conditions. Composting involves the decomposition of substrates rich in organic matter under the influence of microorganisms (Aguilar-Paredes et al. 2023). Among the most important microorganisms involved in the process are bacteria, fungi and actinomycetes. By providing organic matter and oxygen, these organisms successively decompose the feedstock. The result of this process will be compost—a fertiliser rich in organic matter. The composting process is based on the phenomena of mineralisation and humification. As mineralisation proceeds, some organic compounds are converted to mineral forms, i.e. those readily available to plants. On the other hand, the persistence of the process over time allows the humification process, i.e. the formation of complex humus compounds in the composted mixture (Serra-Wittling et al. 1996).

The final product of the composting process will be compost, which is a stabilised organic fertiliser. Due to its properties, it can be successfully used in agriculture (Noor et al. 2023). Compost not only provides the minerals that plants need for growth and development but also organic matter (Felipó 1996). The use of compost improves both soil fertility and soil structure or water and air conditions. Among the main barriers and limitations of the composting process is the possibility of heavy metals or pathogenic organisms (Li et al. 2022). Heavy metals can occur as a result of the use of certain substrates, e.g. sewage sludge. Pathogens present in the substrates will mostly be annihilated during the composting process, but high temperatures during the process are necessary (Karimi et al. 2017). Attention should also be paid to the possibility of odours emanating from the composted heap. This argument is one of the most frequently raised as a public concern, in waste management processes. For this reason, care should be taken both to ensure the quality of the substrates used in the process and to control the process. A schematic of the composting process is shown in Fig. 3.1.

Fig. 3.1 Composting process

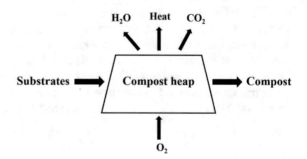

Stages of the Composting Process

Composting is the process of biodegradation of organic matter under aerobic conditions. To a certain field, it represents the phenomena occurring in the soil environment after the supply of organic matter. The composting process is based on two main phenomena. The first is the decomposition of organic matter and its mineralisation. The second is the humification process, which takes longer than mineralisation.

Composting, as well as anaerobic digestion, is a complex process. Both processes occur in four stages, but they are different. Within composting, four phases can be identified (Zhan et al. 2023).

(1) Mesophilic phase—during this phase, the composting process is initiated and the multiplication of microorganisms. The microorganisms begin to decompose organic matter such as simple sugars, proteins or fats into simpler compounds.
(2) Thermophilic phase—during this phase, the decomposition of organic matter by microorganisms takes place to an increasing degree, and an intensive composting process takes place. This contributes to an increase in temperature, even around 70–80 °C. During the thermophilic phase, most of the readily available organic compounds are decomposed into carbon dioxide and converted into humus.
(3) Cooling phase—during this phase, there is a reduction in the number of microorganisms involved in the process, which is related to the depletion of readily decomposable compounds. The effect of the reduction in the number of microorganisms and their activity is a drop in the temperature of the composted mixture, even below 30 °C. Substances that are difficult to decompose, such as lignin, remain in the mixture.
(4) Maturation phase—during this phase, there is an increase in the humus fraction. The conditions in the compost heap stabilise and the temperature of the compost drops to the ambient temperature.

Composting Process Parameters

Both anaerobic digestion and composting, which are biological waste treatment processes, are complex. They are also not easy to manage. This is due, among other things, to the need to provide the microorganisms with the right conditions for growth and development (Lin et al. 2018). The ability to carry out the process and its speed will depend on this. A key role is related to providing the microorganisms with the right conditions. This relates to both the substrates supplied to the process and the process parameters. As in the case of anaerobic digestion, several parameters need to be taken into consideration in composting, in order to ensure the right process conditions. As both processes are classified as biological and based on the use of microorganisms, the key parameters for the composting process will be similar to those for anaerobic digestion. However, their values will be different. Among the most important parameters for the methane digestion process are: presence of oxygen, temperature of the composted mixture, dry matter, C/N ratio and pH. Selected parameters of the substrates used in the composting process are discussed in Table 3.1.

Table 3.1 Selected parameters of substrates used in the composting process

Properties of substrates	Description
Chemical composition	The chemical composition of the substrates used in the composting process is the most important aspect influencing their usability. It also greatly influences the properties of the compost produced. Feedstock for the composting process is a source of both organic matter and the macronutrients and micronutrients necessary for organisms to grow and development
Dry matter	The dry matter content indicates the percentage of solids in the substrate analysed. The higher the value, the drier the substrate. The unit of dry matter content is %
Organic matter content	The organic matter content is the amount of solids in the residue lost during its combustion at 550 °C

As mentioned earlier, composting is a complex biological process, even though the process and its monitoring seems to be simpler than in the case of anaerobic digestion. In order to ensure that the process runs correctly, resulting in the production of compost of the desired quality, selected process parameters must be systematically analysed and optimised. Among the most important parameters of the composting process are: presence of oxygen, dry matter of feedstock, temperature of the composted substrate mixture, C/N ratio and pH. It is particularly important to analyse the adequate aeration of the heap. Too little air can lead to anaerobic conditions. Too much air, on the other hand, can lead to cooling of the composted mixture, including stopping the process. Although composting appears to be a simpler way of treating waste and by-products than fermentation, care must be taken to ensure that it is carried out correctly. Selected parameters related to the composting process, together with a brief characterisation, are presented in Table 3.2.

Technologies of the Composting Process

Composting can be carried out in many ways (Mengistu et al. 2018). The fundamental choice usually concerns the composting place. In this case, a distinction can be made between composting in open heaps and composting in reactors. The choice of technology is usually determined by the financial possibilities for the construction of the plant and the type of substrates used (Wang et al. 2023).

Biomass composting on heaps is the oldest method. This technology has been used for many years all over the world, due to its simplicity and minimal cost. The initial step is to fragment all substrates. This will increase the microorganisms' access to organic matter. The substrates are then mixed in the correct proportions. The next step is the formation of the compost heaps, usually in trapezoidal cross-section. The width of the heaps is adapted to the operating ranges of the compost turning equipment, and the length can reach up to several tens of metres. In the simplest cases, when composting is of an individual nature for a household or local community, a smaller heap can be formed and turned without the use of specialised equipment. Alves et al. (2023) conducted research on community composting strategies for bio-waste treatment. It is extremely important to provide the composted mixture

Table 3.2 Selected parameters related to the composting process

Parameters of the anaerobic digestion process	Description
Oxygen presence	Composting is an aerobic process. The organisms involved in the composting process require the presence of oxygen at all times. As a result, composting can be carried out both in reactors and in open heaps. This is a great advantage and convenience. Irrespective of the choice of technology, however, it is important to ensure sufficient oxygen through aeration
Feedstock dry matter	The dry matter content of the composted mixture can vary depending on the technology used. Regarding the optimum moisture content of the feedstock composting mixture, opinions are divided. It also depends on the type of feedstock and its susceptibility to water release. In many cases, however, a dry matter content of 40–60% will be adequate
Temperature	Temperature is an important parameter related to the composting process. The starting temperature of the composting process is directly related to the ambient temperature. As the composting process continues, the temperature changes. Maximum values even above 60 °C occur in the thermophilic phase
C/N ratio	Each substrate has its own characteristic content of specific elements. In order to provide the microorganisms that carry out the composting process with conditions for growth, care must be taken to ensure an appropriate C/N ratio. According to Azim et al. (2018), following Choi (1999), the optimal C/N ratios in composting of most materials have been reported to vary from 25 to 30
pH	The pH is one of the most important parameters affecting the composting process. The value of this parameter is influenced by the individual substrates that form the feedstock. As reported by Azim et al. (2018), organic material can be composted in a broad pH, and optimal values are between 5.5 and 8. However, it should be noted that the pH value changes during the composting process

with the air necessary for the composting process. For this reason, a distinction can be made between static heaps (with or without aeration) and those in which the compost will be flipped. Under the simplest conditions, the composting process of the compost heap takes place in an open space. This can involve susceptibility to weather conditions. Rain or wind could cool the temperature of the composted mixture. It is therefore advisable to cover the compost heaps. The simplest solution is a roof. Specialised membranes are also increasingly being used. With these, air can also be captured and purified before being released into the atmosphere (Al-Alawi et al. 2019). Gaseous emissions associated with the composting process include CO_2, H_2O and NH_3 (Czekała et al. 2022a, b).

The second way of conducting the composting process is reactor composting, otherwise known as container composting (Czekała et al. 2017). This process takes place in sealed reactors, which can have a capacity of up to several tens of m^3, for each chamber. Air is introduced into the reactors as required. With a closed system, not only can the oxygen content entering the chamber be controlled, but many other parameters can be analysed (Liu et al. 2020). The most important of these includes temperature or gaseous emissions from the composted mixture. Another advantage is the ability to collect leachate and condensate emitted from the mixture. Also, deodorisation will be more effective than with open heaps. This is due to the possibility of using biofilters. Because conditions can be controlled to a very high level, the composting process in reactors is shorter. In practice, composting in a reactor may last only a dozen days or so, after which the compost must be matured. Maturing can take place outside the chambers, e.g. in the yard, to allow the next batch of feedstock to undergo the composting process in the reactors. Figure 3.2 shows the chambers where the composting process takes place.

A specific type of composting process is vermicomposting (Ducasse et al. 2022). This is a composting process with the participation of earthworms. Thanks to their presence, the process can take place faster, and the compost produced will be of better quality. However, care must be taken to ensure that conditions are preferred by both the earthworms and the other microorganisms that process organic matter. The process will result in compost with highly fertilising properties. This product is also characterised by its ability to store water. Before it can be used in agriculture or horticulture, the earthworms must be separated and transferred to a new batch of composted substrates. A paper by Garg and Gupta (2009) discusses the possibility

Fig. 3.2 Chambers in which the composting process takes place

of using agro-industrial processing waste in vermicomposting. Substrates such as sugar industrial waste, winery waste, crop residues, coffee pulp, oil industry waste, food and industry waste, among others, were analysed.

Advantages and Disadvantages of Composting
Composting is an alternative waste management method to anaerobic digestion. Both processes can manage numerous biodegradable waste, including waste and by-products from food industry. Both processes have advantages as well as disadvantages or limitations. Often the two processes are compared with each other (Murphy and Power 2012). The undoubted characteristic is the ability to convert organic waste into a valuable fertiliser. For aerobic conditions, this will be compost, and for anaerobic conditions, it will be digestate for wet technology or digestion residues for dry technology. This will be directly related to the reduction in mass and volume, relative to the volumes characterising the initial condition (Czekała 2019). The main difference will be the potential for energy production. In the anaerobic digestion process, biogas is produced. In the case of composting, it is possible to recover energy in the form of heat (Fan et al. 2021), although this solution is not often used. This aspect is often crucial when deciding on an installation. It is related to the costs, which are much lower for composting than is the case for biogas plants. Composting is a simpler process to carry out and to optimise. Even in the case of technological problems or failures, the intensity of the process can be reduced for a certain period of time, even including the opening of the reactors. In the case of biogas production, this is not possible, due to the inability of the fermented waste to be oxygenated. Another advantage of composting is the greater reduction in waste mass and volume, compared to the same raw materials undergoing anaerobic digestion. For composting, the mass reduction is most often between 30 and 50%, and can even exceed 50%. For decomposition under anaerobic conditions, the reduction is most often up to 30%.

3.3 Waste and By-Product from Food Industry as Substrates for Composting Process

The generation of waste and by-products is an integral process associated with the functioning of the food industry. The processing of all primary raw materials is characterised by the generation of both main and by-products and waste (Khedkar and Singh 2018). As the amount of waste and by-products generated in the plants concerned can be estimated relatively accurately, the possibility of their utilisation arises. In addition to their management for feed, fertiliser or energy purposes, much attention is being paid to the use of these waste as a substrate for composting.

In addition to providing suitable substrates rich in organic matter and an optimum C/N ratio and pH, other factors must also be taken care of. Since composting takes place under aerobic conditions, it is necessary not only to provide air, but also to ensure that air can circulate through the composted mixture. Before composting, the waste is shredded with a mobile shredder if necessary, or with a chipper in the case of shredding waste from parks and gardens, and then fractionated on a drum sieve.

The vast majority of waste and by-products from food industry can be used to produce compost. A high organic matter content and a relatively high water content are factors that make these substrates desirable for microorganisms. By selecting suitable waste mixtures, it is possible not only to manage the waste, but also to prepare compost with suitable parameters that will be an attractive product for use as fertiliser (Shilev et al. 2007). The benefits of using waste and by-products from food industry in aerobic biodegradation can be analysed on several levels. Of particular note is their role as a source of organic matter and nutrients and their use as structure-improving substrates—bulking agents (Malińska and Zabochnicka-Świątek 2013).

Source of Organic Matter

According to researchers, composting is considered as environmentally friendly method, and compost can be used as a fertiliser (Waqas et al. 2023). Public acceptability for composting is often higher than that for anaerobic digestion. The process of composting is directly linked to changes in organic matter (Kulikowska and Klimiuk 2011). As the composting process continues, organic matter decomposes, and the compost produced is rich in valuable organic compounds (Sugahara and Inoko 1981). Ghinea and Leahu (2020) conducted a study on the composting process using such food industry waste as apple, banana, orange, kiwi peels, cabbage leaves, potato and carrot peels. These were therefore waste that provided both a rich source of organic matter and the nutrients. According to the authors, compost was successfully produced and meets the requirement standards for agricultural use. One of the most popular raw materials used for composting is straw (Piao et al. 2023). Its special role is related both to its organic matter content and to its role as a bulking agent.

Source of Nutrients

Composting is a biomass biodegradation process that can take place for a single substrate as well as for several co-substrates used in a mixture. An example of a substrate that can be composted without additives is manure. However, it is far more common to compost several substrates together. This is due to, among other things, the possibility of reducing gaseous emissions (Barthod et al. 2018) or the desire to manage a variety of biomass. Another reason will be to create conditions that allow the composting process to proceed properly. An example would be composting sewage sludge together with straw. The first waste is abundant in nitrogen and the second in carbon. This makes it relatively easy to select a suitable C/N ratio (Czekała et al. 2016). Straw can be substituted by some feedstocks from the food industry, e.g. brewers' malt. The authors pointed out that the addition of carbon-rich by-products can enhance the optimisation of the C/N ratio and pH. This also reduces the cost of straw expenses. Another type of waste from food industry used in the composting process is bakery waste consisting of cake skins, cake waste and expired breads and cakes (Govindaraju et al. 2021). The authors concluded that the compost could potentially be used as an organic amendment for crops in the agricultural field.

Bulking Agent

A special role is played by structuring substrates that allow the supply of oxygen throughout the whole volume of the heap/reactor. On the one hand, this ensures favourable conditions for composting; on the other hand, it reduces the possibility of anaerobic conditions. Among the bulking agents are wood chips, wheat straw, sawdust, rice husk, rice bran, chopped hay, wood shavings and peanut shells (Chang and Chen 2010). Among bulking agents, coco peat, for example, is less used worldwide, which is in turn popular in South Korea (Lee et al. 2020).

The porous structure ensures that oxygen can circulate, allowing the process to take place properly throughout the composted mixture. Substrates that act as bulking agents may sometimes require shredding. Examples include straw or hay, which are relatively easy to grind, even using mobile equipment. Other raw materials, such as dried pomace or peelings, will not require shredding.

The main aim of the composting process is to produce a high-quality and environmentally safe fertiliser. The properties of the produced product will depend, among other things, on the substrates used and the process technology. By using substrates of known origin, free of impurities, a high-quality, stabilised fertiliser can be produced. An example of a fertiliser produced at one installation in Switzerland is shown in Fig. 3.3. Processing waste and by-products from food industry can be a good waste management solution, especially for local companies.

Fig. 3.3 High-quality compost for commercial retail (Switzerland)

3.4 Conclusion

The composting process is associated with a number of advantages that contribute to its popularity. Composting is a relatively easy process to carry out, and the capital expenditure is not as high as for anaerobic digestion. This is especially the case when considering heap composting. The construction and operating costs of such a plant are relatively small, and sometimes even two people are sufficient to operate it. However, it may become necessary to use more complex solutions to reduce emissions into the environment due to environmental guidelines. This applies particularly to gaseous emissions. The composting process mainly emits CO_2 and water vapour, but ammonia, hydrogen sulphide and other compounds, including odours, can also be emitted. The number of installations treating a variety of waste by composting is steadily increasing. This includes sludge, bio-waste and separated organic fractions from the mixed municipal waste stream (mechanical–biological treatment facilities for municipal waste). This will, however, have an impact on the increased environmental requirements for the approval of installations.

In many cases, waste and by-products from food industry are excellent substrates for the composting process. The large number of modern composting technologies and their variety make it possible to select one that is suitable for the waste to be processed and the financial possibilities of the investor. If the process is carried out correctly and the product is of good quality, it is possible to apply for certification of the compost as an organic fertiliser. In this case, benefits can be achieved both for accepting the waste and for selling valuable compost, especially since compost can have a wide range of applications. In addition to agriculture, it is used, among other things, in forestry, in fertilising green spaces or in the reclamation of degraded land. Thanks to its properties, it is able to restore soil fertility to a certain extent while reducing the use of artificial fertilisers. Thus, it is possible to reduce crop production costs while limiting the negative environmental impact of artificial fertilisers.

References

A. Aguilar-Paredes, G. Valdés, N. Araneda, E. Valdebenito, F. Hansen, M. Nuti, Microbial community in the composting process and its positive impact on the soil biota in sustainable agriculture. Agronomy **13**, 542 (2023). https://doi.org/10.3390/agronomy13020542

M. Al-Alawi, T. Szegi, L. El Fels, M. Hafidi, B. Simon, M. Gulyas, Green waste composting under GORE(R) cover membrane at industrial scale: physico-chemical properties and spectroscopic assessment. Int. J. Recycl. Org. Waste Agric. **8**(Suppl 1), 385–397 (2019). https://doi.org/10.1007/s40093-019-00311-w

D. Alves, V.I. Mato, S, Community composting strategies for biowaste treatment: methodology, bulking agent and compost quality. Environ. Sci. Pollut. Res. (2023). https://doi.org/10.1007/s11356-023-25564-x

K. Azim, B. Soudi, S. Boukhari, C. Perissol, S. Roussos, I. Thami Alami, Composting parameters and compost quality: a literature review. Org. Agr. **8**, 141–158 (2018). https://doi.org/10.1007/s13165-017-0180-z

J. Barthod, C. Rumpel, M.F. Dignac, Composting with additives to improve organic amendments. A review. Agron. Sustain. Dev. **38**, 17 (2018). https://doi.org/10.1007/s13593-018-0491-9

W. Bojarski, W. Czekała, M. Nowak, J. Dach, Production of compost from logging residues. Bioresour. Technol. **376**, 128878 (2023). https://doi.org/10.1016/j.biortech.2023.128878

J.I. Chang, Y.J. Chen, Effects of bulking agents on food waste composting. Bioresour. Technol. **101**(15), 5917–5924 (2010). https://doi.org/10.1016/j.biortech.2010.02.042

K. Choi, Optimal operating parameters in the composting of swine manure with wastepaper. J. Environ. Sci. Health **34**(6), 975–987 (1999). https://doi.org/10.1080/03601239909373240

W. Czekała, J. Dach, D. Janczak, A. Smurzyńska, A. Kwiatkowska, K. Kozłowski, Influence of maize straw content with sewage sludge on composting process. J. Water Land Dev. **30**, 43–49 (2016). https://doi.org/10.1515/jwld-2016-0020

J. Czekała, R. Dach, D. Dong, K. Janczak, K. Malińska, K. Jóźwiakowski, A. Smurzyńska, M. Cieślik, Composting potential of the solid fraction of digested pulp produced by a biogas plant. Biosyst. Eng. **160**, 25–29 (2017). https://doi.org/10.1016/j.biosystemseng.2017.05.003

W. Czekała, P.P. Janczak, M. Nowak, J. Dach, Gases emissions during composting process of agri-food industry waste. Appl. Sci. **12**(18), 9245 (2022a). https://doi.org/10.3390/app12189245

W. Czekała, D. Janczak, P. Pochwatka, M. Nowak, J. Dach, Gases emissions during composting process of agri-food industry waste. Appl. Sci. **12**, 9245 (2022b). https://doi.org/10.3390/app12189245

W. Czekała, M. Nowak, G. Piechota, Sustainable management and recycling of anaerobic digestate solid fraction by composting: a review. Bioresour. Technol. **375**, 128813 (2023). https://doi.org/10.1016/j.biortech.2023.128813

W. Czekała, Biogas production from raw digestate and its fraction. J. Ecol. Eng. **20**(6), 97–102 (2019). https://doi.org/10.12911/22998993/108653

V. Ducasse, Y. Capowiez, J. Peigné, Vermicomposting of municipal solid waste as a possible lever for the development of sustainable agriculture. A review. Agron. Sustain. Dev. **42**, 89 (2022). https://doi.org/10.1007/s13593-022-00819-y

S. Fan, A. Li, A. ter Heijne, C.J.N. Buisman, W.-S. Chen, Heat potential, generation, recovery and utilization from composting: a review. Resour. Conserv. Recycl. **175**, 105850 (2021). https://doi.org/10.1016/j.resconrec.2021.105850

M.T. Felipó, Compost as a source of organic matter in mediterranean soils, in *The Science of Composting*, ed. by M. de Bertoldi, P. Sequi, B. Lemmes, T. Papi (Springer, Dordrecht, 1996). https://doi.org/10.1007/978-94-009-1569-5_38

V. Garg, R. Gupta, Vermicomposting of agro-industrial processing waste, in *Biotechnology for Agro-Industrial Residues Utilisation*, ed. by P. Singh nee' Nigam, A. Pandey (Springer, Dordrecht, 2009). https://doi.org/10.1007/978-1-4020-9942-7_24

C. Ghinea, A. Leahu, Monitoring of fruit and vegetable waste composting process: relationship between microorganisms and physico-chemical parameters. Processes **8**, 302 (2020). https://doi.org/10.3390/pr8030302

M. Govindaraju, K.V. Sathasivam, K. Marimuthu, Waste to wealth: value recovery from bakery wastes. Sustainability **13**, 2835 (2021). https://doi.org/10.3390/su13052835

H. Karimi, M. Mokhtari, F. Salehi, S. Sojoudi, A. Ebrahimi, Changes in microbial pathogen dynamics during vermicomposting mixture of cow manure–organic solid waste and cow manure–sewage sludge. Int. J. Recycl. Org. Waste Agric. **6**, 57–61 (2017). https://doi.org/10.1007/s40093-016-0152-4

R. Khedkar, K. Singh, Food industry waste: a panacea or pollution hazard? in *Paradigms in Pollution Prevention. SpringerBriefs in Environmental Science*, ed. by T. Jindal (Springer, Cham, 2018). https://doi.org/10.1007/978-3-319-58415-7_3

D. Kulikowska, E. Klimiuk, Organic matter transformations and kinetics during sewage sludge composting in a two-stage system. Bioresour. Technol. **102**, 10951–10958 (2011). https://doi.org/10.1016/j.biortech.2011.09.009

J.H. Lee, D. Luyima, C.H. Lee, S.-J. Park, T.-K. Oh, Efficiencies of unconventional bulking agents in composting food waste in Korea. Appl. Biol. Chem. **63**, 68 (2020). https://doi.org/10.1186/s13765-020-00554-6

M. Li, G. Song, R. Liu, X. Huang, H. Liu, Inactivation and risk control of pathogenic microorganisms in municipal sludge treatment: a review. Front. Environ. Sci. Eng. **16**, 70 (2022). https://doi.org/10.1007/s11783-021-1504-5

L. Lin, F. Xu, X. Ge, Y. Li, Improving the sustainability of organic waste management practices in the food-energy-water nexus: A comparative review of anaerobic digestion and composting. Renew. Sust. Energ. Rev. **89**, 151–167 (2018). https://doi.org/10.1016/j.rser.2018.03.025

Z. Liu, X. Wang, F. Wang, Z. Bai, D.R. Chadwick, T. Misselbrook, L. Ma, The progress of composting technologies from static heap to intelligent reactor: Benefits and limitations. J. Clean. Prod. **270**, 122328 (2020)

K. Malińska, M. Zabochnicka-Świątek, Selection of bulking agents for composting of sewage sludge. Environ. Prot. Eng. **39**(2), 91–103 (2013). https://doi.org/10.5277/EPE130209

T. Mengistu, H. Gebrekidan, K. Kibret, K. Woldetsadik, B. Shimelis, H. Yadav, Comparative effectiveness of different composting methods on the stabilization, maturation and sanitization of municipal organic solid wastes and dried faecal sludge mixtures. Environ. Syst. Res. **6**, 5 (2018). https://doi.org/10.1186/s40068-017-0079-4

V. Mozhiarasi, T.S. Natarajan, Slaughterhouse and poultry wastes: management practices, feedstocks for renewable energy production, and recovery of value added products. Biomass Conv. Bioref. (2022). https://doi.org/10.1007/s13399-022-02352-0

J.D. Murphy, N.M. Power, A technical, economic and environmental comparison of composting and anaerobic digestion of biodegradable municipal waste. J. Environ. Sci. Health Tox. Hazard Subst. Environ. Eng. **41**, 865–879 (2012). https://doi.org/10.1080/10934520600614488

R.S. Noor, F. Hussain, I. Abbas, M. Umair, Y. Sun, Effect of compost and chemical fertilizer application on soil physical properties and productivity of sesame (*Sesamum Indicum* L.). Biomass Conv. Bioref. **13**, 905–915 (2023). https://doi.org/10.1007/s13399-020-01066-5

M. Piao, A. Li, H. Du, Y. Sun, H. Du, H. Teng, A review of additives use in straw composting. Environ. Sci. Pollut. Res. **30**, 57253–57270 (2023). https://doi.org/10.1007/s11356-023-26245-5

C. Serra-Wittling, E. Barriuso, S. Houot, Impact of composting type on composts organic matter characteristics, in *The Science of Composting*, ed. by M. de Bertoldi, P. Sequi, B. Lemmes, T. Papi (Springer, Dordrecht, 1996). https://doi.org/10.1007/978-94-009-1569-5_26

A. Sharma, R. Soni, S.K. Soni, Decentralized in-vessel composting: an efficient technology for biodegradable solid waste management. Biomass Conv. Bioref. (2023). https://doi.org/10.1007/s13399-023-04508-y

S. Shilev, M. Naydenov, V. Vancheva, A. Aladjadjiyan, Composting of food and agricultural wastes, in *Utilization of By-Products and Treatment of Waste in the Food Industry*, vol. 3, ed. by V. Oreopoulou, W. Russ (Springer, Boston, MA, 2007). https://doi.org/10.1007/978-0-387-35766-9_15

K. Sugahara, A. Inoko, Composition analysis of humus and characterization of humic acid obtained from city refuse compost. Soil Sci. Plant Nutr. **27**, 213–224 (1981)

H. Wang, Y. Qin, L. Xin, C. Zhao, Z. Ma, J. Hu, W. Wu, Preliminary techno-economic analysis of three typical decentralized composting technologies treating rural kitchen waste: a case study in China. Front. Environ. Sci. Eng. **17**, 47 (2023). https://doi.org/10.1007/s11783-023-1647-7

M. Waqas, S. Hashim, U.W. Humphries, S. Ahmad, R. Noor, M. Shoaib, A. Naseem, P.T. Hlaing, H.A. Lin, Composting processes for agricultural waste management: a comprehensive review. Processes **11**, 731 (2023). https://doi.org/10.3390/pr11030731

Y. Zhan, Y. Chang, Y. Tao, H. Zhang, Y. Lin, J. Deng, T. Ma, G. Ding, Y. Wei, J. Li, Insight into the dynamic microbial community and core bacteria in composting from different sources by advanced bioinformatics methods. Environ. Sci. Pollut. Res. **30**, 8956–8966 (2023). https://doi.org/10.1007/s11356-022-20388-7

X. Zhou, J. Yang, S. Xu, J. Wang, Q. Zhou, Y. Li, X. Tong, Rapid in-situ composting of household food waste. Process. Saf. Environ. Prot. **141**, 259–266 (2020). https://doi.org/10.1016/j.psep.2020.05.039

Chapter 4
Environmental Aspect of Waste and By-Product from Food Industry and Their Management

4.1 Introduction

Environmental issues are becoming more important every year (Mwanza and Mbohwa 2022). This is due to increasingly stringent requirements in some countries, as well as concerns about the climate change. Climate change has affected the whole world, and rapid action is required. Extreme climatic conditions, including droughts, heat waves, heavy rains, floods and landslides, are becoming more and more frequent in the world (Kameni Nematchoua and Orosa 2023).

People are becoming increasingly aware of the negative impact of anthropogenic changes on the environment. Waste management plays a special role in this effort. This is because of that waste to have a negative impact on every element of the biosphere. On the other hand, if handled rationally, waste can turn out to be a valuable raw material, and its recycling will save the acquisition of raw materials. Such action will be in line with the principle of circular economy (Kabir and Kabir 2022).

Biodegradable waste which can decompose under almost any conditions (Czekała et al. 2023) is relatively large. Decomposition is possible under both aerobic and anaerobic conditions, which is particularly important in terms of environmental risk and protection. This is because uncontrolled decomposition can cause the emission of harmful compounds into the atmosphere, pedosphere and hydrosphere (Siddiqua et al. 2022). The biosphere may also suffer due to contact with harmful substances. This is why it is so important to take action to eliminate all kinds of hazards.

© The Author(s), under exclusive license to Springer Nature Switzerland AG 2023
W. Czekała, *Biological Treatment of Waste and By-Products from Food Industry*,
SpringerBriefs in Applied Sciences and Technology,
https://doi.org/10.1007/978-3-031-47487-3_4

4.2 Environmental Aspect of Waste and By-Product from Food Industry and Their Management

The increase in the earth's population, the depletion of renewable resources, the limited available land for cultivation or waste management problems are just some of the factors strongly affecting food production and the management of waste and by-products from food industry. Torres-León et al. (2018) reported that one possibility for the management of waste and by-products is their use in the food sector. The authors point out that biomaterials have ample potential for generating food additives, which could minimise malnutrition and hunger in the developing countries where it is produced. This would be rational due to local use. In many cases, however, the waste and by-products from food industry require other uses.

Management options for the waste in question include fertiliser use, thermal treatment and biological methods including anaerobic digestion and composting. A comparison of selected parameters characterising anaerobic digestion and composting is presented in Table 4.1.

The use of waste and by-products from food industry in composting and anaerobic digestion processes is associated with numerous requirements. This includes both regulations and approaches to their management (Shen et al. 2023). Biodegradable waste requires a special approach in its management. Unlike most waste such as plastic, glass or construction waste, biodegradable waste needs to be managed as quickly as possible. As noted by Kaur et al. (2023), treatment of food waste has become a need of the hour. The reason is to prevent degradation of biodiversity, but also to recover the essential nutrients and other value-added stuffs from food waste.

Waste should be stored for the shortest time possible. Under unfavourable storage conditions, e.g. high temperatures, it only takes a few hours for the substrates to start

Table 4.1 Comparison of selected parameters characterising anaerobic digestion and composting

Parameter	Anaerobic digestion	Composting
Substrates	Most biodegradable substrates can be used in both processes	
Product/products	Biogas/biomethane Digestate	Compost
Mass and volume reduction (%)	Less than composting	Higher than anaerobic digestion
Energy	Obtaining biogas that can be converted into biomethane or electricity and heat in cogeneration	Possibility of heat recovery
Gaseous emissions	The process takes place in sealed reactors	Gaseous emissions from heap composting
Cost of investment	Higher than composting	Lower than anaerobic digestion

decomposing. If unfavourable conditions occur, not only will uncontrolled decomposition and emissions occur, but the energy potential of these waste will be partially lost.

It should be mentioned that when decomposed, the waste will lose its potential to produce biogas, or compost, due to the initiation of decomposition processes. Therefore, it becomes necessary to manage biodegradable waste in a rational manner that will allow it to be treated as part of the recovery process, with the lowest possible environmental impact (Obuobi et al. 2023). It is also important to remember to separate waste selectively and not to divert contaminated waste to the installation. This will require additional treatment, and the result in terms of raw material quality will not always be satisfactory. The environmental impact of waste and by-product from food industry is often discussed with a division into pedosphere, hydrosphere, atmosphere, biosphere and anthroposphere.

The use of waste with contaminants can result in them entering to the ecosystem. One case could be inadequate storage, resulting in leachate entering the soil environment (Fig. 4.1). This is particularly important in the case of fertiliser use of waste, a trend that is becoming increasingly popular (Siddiqui et al. 2023). O'Connor et al. (2022) analysed current and emerging contaminants in food waste and their associated impacts on environmental, soil and human health. The authors highlighted the role of, among other things, the ability to bioaccumulate and biomagnify contaminants.

Fig. 4.1 Emissions of leachate from manure heap

The hydrosphere is also exposed to the negative impacts of waste and by-products. The main possibility of water pollution is related to the illegal dumping of wastewater into water reservoirs. Another type of hazard is the introduction of waste leachate into groundwater. There are cases of food-related waste being dumped into sewers. It has also been highlighted that the introduction of food waste disposers will have a negative effect on maintaining the function of municipal wastewater treatment plants and reducing greenhouse gases (Kim and Phae 2023). Therefore, it should not be used. Both manure and waste by-products from food industry can be a source of leachate emissions to the environment.

One of the easiest ways to dispose of waste is to burn it. In some countries, it is common for waste such as leaves, grass clippings, plant stalks, vines, weeds, twigs and branches to be burned (Hashim et al. 2022). Elsewhere in the world, local bans on thermal waste conversion are in place. This is primarily related to restrictions on the potential for atmospheric pollution. As reported by Piao et al. (2023), composting provides a method to combat air pollution from straw burning. This is another important aspect in favour of using waste and by-products from food industry in biological processes. Uncontrolled biodegradation of waste, including that from the food industry, is a source of gaseous pollutants such as methane, carbon dioxide, ammonia and hydrogen sulphide. It is therefore advisable to manage this waste using techniques that reduce the negative impact on the environment. One example would be waste treatment in biogas plants. Waste decomposition is often a source of odours. According to Tamburini et al. (2020), the use of anaerobic digestion can reduce odour impact by more than 90%. This will depend primarily on the types of substrates used.

The production and harvesting of food have a significant impact on the environment, encompassing both the use of energy, water and other resources and the impact on the various elements of the planet. The impact on animals and humans is also not insignificant. According to Khedkar and Singh (2018), techniques of industrial waste management can be classified into three options: source reduction (by processing plant modification), waste recovery, recycle or waste treatment for value-added products and the last one eco-friendly detoxification or neutralisation of the undesirable components. Choosing the right way will not only affect environmental protection, but also the costs and revenues associated with the operation of each plant.

There are environmental, social and economic benefits to be gained from treating this waste in biogas and composting plants. The environmental benefits include, in particular, that the waste is recycled and not sent to landfill or other unsuitable locations, such as forests or rivers. Society stands to gain from this improvement in the local environment. By doing so, waste will be a source of energy instead of being stored in uncontrolled conditions. This is especially important in areas where there are problems with access to energy, e.g. on islands or in areas away from major cities. In this case, the biogas plant could turn out to be a power station supplying not only the food industry, but also the local community. The economic benefits will be visible especially for the production plant. However, it is worth mentioning an important benefit for the local community acquiring heat from cogeneration and valuable digestate on preferential terms. In Poland, for example, some biogas plants give heat away to the community for free or for a small fee, several times lower than heat generated from heating a building with coal. The digestate, on the other hand, can be purchased at a price of a few EUR per 1 m^3, which, compared to prices of several hundred EUR for 1 Mg artificial fertilisers, becomes a very attractive alternative, despite the lower nutrient content of the fertiliser coming from the biogas plant. However, when analysing all the benefits of treating organic waste in both a biogas plant and a composting plant, it has to be said that the operation of both installations has a positive impact on the natural environment, especially in the local aspect. For this reason, these installations have a strong development perspective worldwide.

4.3 Waste and By-Product from Food Industry in Circular Economy Context

Issues related to sustainability and the circular economy are becoming more important every year. Awareness of resource constraints, rising energy prices or global climate change are just some of the issues facing humanity in the twenty-first century. Survival will depend on how people deal with the problems (Guerra et al. 2022). The circular economy is an increasingly implemented measure by many countries around the world. This is due to both a rational approach to current problems and cost savings. This is because, in many cases, recycling and recovery are economically justifiable processes in relation to the production of new products using raw materials. In addition, the fact that more and more cities are working towards becoming smart cities and being sustainable (Yigitcanlar et al. 2019) adds stature to the situation. It is therefore imperative to seek and implement solutions that fit into the trend.

The food industry is one of the largest consumers of energy, water and raw materials. Reductions in the energy, fuels, water and raw materials consumption are a key topic in terms of the profitability of business operations. Practices undertaken in this area are also becoming increasingly important for society (Compton et al. 2018). Waste can be a tool that can be helpful in working with sustainability in the food industry. In the past, waste was seen as a problem. Currently, they are increasingly seen as a valuable resource. Using waste for both biogas and compost production reduces the use of fossil fuels and fertilisers.

The initial actions taken by many countries were based on a linear model. This model was characterised by maximum acquisition of raw materials and their use. This also resulted in an increase in the amount of waste generated at each stage of processing (Jørgensen and Pedersen 2018). In a circular economy system, the waste generated is used further in production processes for as long as it is possible and environmentally justified. As a result, their lifetime is longer. In the circular economy, the waste and by-products generated in the production cycle are used as raw materials for further processes. This reduces the amount of waste, as well as the amount of raw materials obtained. The food industry and similar economic sectors are places where these solutions can be implemented (Lahiri et al. 2023).

Activities in line with the circular economy are based on reducing the use of raw materials as much as possible. In addition, it is important to reduce energy and water consumption. This is extremely important because the processing of raw materials is associated with high consumption of both water and energy (Czekała 2021). Also important is the introduction of emissions into the environment, which, according to the circular economy principle, should be kept to a minimum. Waste and by-product from food industry can also take part in all these activities.

The raw materials in question can be much needed for both anaerobic digestion and composting. Each installation has a characteristic capacity. This is the amount of raw material with specific parameters that needs to be delivered over a given period of time (e.g. a day, a month or a year) in order for the installation to operate at maximum efficiency. Ideally, this value should be close to 100%. This means that the efficiency of the installation has been used to the maximum. There are cases where even less than 50% is used, which has a negative impact on the balance of operation of the investment.

An example of a solution to achieve higher profits is to process waste instead of substrates from, e.g. targeted crops. For example, the acquisition of maize silage for the anaerobic digestion process is associated with a cost. The same is for example, for sawdust, which is a popular and valuable substrate for composting. It is possible to find such waste and other raw materials, the cost of which will be much less, or the acquisition of which will be for free (Czekała 2022a). In specific cases, it is even possible to charge a fee for waste management. One of the best cases is the processing of waste from the meat industry into biogas. For 1 Mg of development of such waste, up to several hundred EUR can be obtained, which is the revenue of the installation.

The key in rational waste management is to avoid or reduce its production. These solutions certainly have the least impact on the environment. However, for some processes, it is not possible to avoid waste generation. This is due to the fact that waste associated with the food industry are generated every day everywhere in the world (Mokrane, et al. 2023), which is related to the need to consume food. If waste and by-product from food industry cannot be avoided or reduced, then it should be used as much as possible, e.g. for energy production. Bancal and Ray (2022) highlighted that food loss and waste in relation to fruits and vegetables are important in terms of both food security and environmental protection. Therefore, the generation of such waste should be reduced, as shown in Fig. 4.2.

Food Loss and Food Waste

One-third of food intentionally grown for human consumption is never consumed and is wasted (Garcia-Garcia et al. 2017). This problem is much broader and should be analysed in an environmental, social and economic context, among others. As reported by Rashwan et al. (2023), food loss and food waste are a major issue affecting many areas. The authors mention: food security, environmental pollution, producer profitability, consumer prices and climate change.

Fig. 4.2 Food waste

Anaerobic digestion is one option for the management of all types of organic waste, both of plant and animal origin (Czekała 2022b). Waste associated with the entire food life cycle plays a special role, including, among others, waste and residues from the agri-food industry or unused and wasted food (Kennard 2020). These substrates are increasingly an alternative to targeted crops and natural fertilisers. This is due to the fact that their properties and diversity allow them to largely replace the aforementioned substrates. It is also important to emphasise their important role in relation to the circular economy on the premises of food production and processing plants, as pointed out by Gonçalves and Maximo (2022), among others.

Food-Energy Conflict
The increasing demand for energy is driving the search for new sources. Among the alternatives to fossil fuels, renewable energy and nuclear energy are mentioned. Despite such an important topic as energy security, there are often opposing voices for renewable energy (Susskind et al. 2022). The food-energy conflict also plays a special role. The production of food such as fruit and vegetables is linked to resources such as land, water, labour and energy (Olatunji et al. 2023). The use of land for the production of agricultural crops for biofuels is controversial (Czekała 2018). Therefore, it is important to look at the possibility of using alternative sources such as food waste and algae. Algae-based biofuels present numerous benefits. They can reproduce very quickly if the right factors are provided, among which are the type of growing system, access to light, nutrients, temperature or pH (Mahmood et al. 2023). However, the use of both waste and by-products from food industry in anaerobic digestion is associated with many challenges (Xu et al. 2018). Considering the numerous benefits, mainly

in the context of energy production, the development and wider application of this direction than before should be pursued.

Fertilisers from Waste

In recent years, the price of artificial fertilisers has increased significantly in many parts of the world (Tröster 2023). This is related, among other things, to the increase in the price of natural gas (Czekała et al. 2022). In the case analysed, waste and by-products from food industry can be used to produce biogas or biomethane having parameters corresponding to the quality of the above-mentioned fossil fuel (Adnan et al. 2019). On the other hand, biogas plant digestate and compost can partially replace the need for artificial fertilisers. It should also be noted that the production and use of artificial fertilisers can be associated with negative environmental impacts, which is highlighted in the context of sustainable agriculture. For this reason, waste materials are increasingly being used to produce fertilisers (Czekała et al. 2019). These are most often organic fertilisers, which, in addition to organic matter, also contain macronutrients and micronutrients in their composition. With this in mind, the availability and use of fertilisers is crucial for the agri-food sector. Their use not only increases efficiency and generally provides higher profits for crop owners, but also promotes an increase in the quantity of food, which is particularly important in the fight against hunger, especially in poor countries.

Water Management

Special attention in the context of the circular economy should also be paid to water management. As is well known, water resources are limited, and there are water deficits in many parts of the world. However, water is essential for human beings, animals and plants. Water is required to produce nearly everything people use and consume, from the food we eat and the clothes we wear to the technological devices that are integral to our modern society (Fulton et al. 2014). The supply of water of adequate quality has always been an important practical issue, as well as for the scientific world. However, analyses of the water footprint are increasingly common, although the topic was already analysed many years ago, for example in the paper Hoekstra and Chapagain (2006).

Mekonnen and Gerbens-Leenes (2020), in their article "The Water Footprint of Global Food Production", published a lot of important information regarding the water footprint. They note that agricultural production is the main water consumer. Increase in demand for water can occur as a result of population growth, income growth and dietary shifts. The authors also highlighted that reducing food waste can have a positive impact on rational water management. As stated by H-Hargitai and Somogyi (2023), water is an ancillary topic in the circularity assessment. However, it should be emphasised that water is essential in many places of food production as well as waste and by-product management from food industry, which is often not noticed. One step towards that goal is to develop and use tools for monitoring circularity that can handle consumer goods and water as raw material beyond durable products.

The last few years have shown how the prices of raw materials, fertilisers or energy can rise even by more than 100% in a year for some regions. For this reason, more and more companies are trying to protect themselves against sudden price changes that might not be realisable for them. As a result, companies are seeking to equip themselves with their own energy production systems. Photovoltaic panels in particular are gaining in importance, but there is also an increased interest in biogas plants. This approach allows a certain degree of independence from market fluctuations. For this reason, more and more companies in the food industry are opting to build biogas plants next to the plant. Furthermore, the reduction in costs associated with the acquisition of raw materials is another benefit for the company. Therefore, more and more attention is being paid to the analysis of specific waste for energy, especially as some of them can be more efficiency than dedicated crops for biogas plants. By using waste to produce biogas and compost, not only economic, but also environmental and social benefits can be achieved.

4.4 Conclusion

Limiting global warming to 1.5 °C and achieving carbon neutrality should be reached by the middle of the twenty-first century. Consequences of climate change include biodiversity loss and sea level rise. Waste management is certainly one of the key areas for sustainable development and climate protection. If managed rationally, it is possible not only to reduce its environmental impact but also to gain economic benefits. Local waste management can both protect the environment and diversify sources of renewable energy. An excellent example is the use of waste and by-products from food industry in biological waste treatment processes, especially in anaerobic digestion. Given the specific properties that characterise waste and by-product from food industry, it is advisable to use these raw materials in biological processes, especially in biogas plants.

Due to the numerous advantages of this energy production technology, biogas plants should be one of the main sources of renewable energy in many countries. Available technologies related to smaller-scale biogas production are becoming more popular not only in developed countries, but also in developing countries. Biogas plants fit positively into the energy policies of many countries around the world, and their operation using waste is definitely in line with the circular economy. However, it is important to ensure that a problem does not arise with the use of good quality food for biofuel production. This topic, together with the competing use of agricultural land for energy crops instead of food production, can be problematic and cause social conflicts.

References

A.I. Adnan, M.Y. Ong, S. Nomanbhay, K.W. Chew, P.L. Show, Technologies for biogas upgrading to biomethane: a review. Bioengineering **6**, 92 (2019). https://doi.org/10.3390/bioengineering6040092

V. Bancal, R.C. Ray, Overview of food loss and waste in fruits and vegetables: from issue to resources, in *Fruits and Vegetable Wastes* (Springer, Singapore, 2022). https://doi.org/10.1007/978-981-16-9527-8_1

M. Compton, S. Willis, B. Rezaie, K. Humes, Food processing industry energy and water consumption in the Pacific northwest. Innov. Food Sci. Emerg. Technol. **47**, 371–383 (2018). https://doi.org/10.1016/j.ifset.2018.04.001

W. Czekała, Solid fraction of digestate from biogas plant as a material for pellets production. Energies **14**(16), 5034 (2021). https://doi.org/10.3390/en14165034

W. Czekała, Digestate as a source of nutrients: nitrogen and its fractions. Water **14**(24), 4067 (2022a). https://doi.org/10.3390/w14244067

W. Czekała, T. Jasiński, M. Grzelak, K. Witaszek, J. Dach, Biogas plant operation: digestate as the valuable product. Energies **15**, 8275 (2022). https://doi.org/10.3390/en15218275

W. Czekała, T. Jasiński, J. Dach, Profitability of the operation of agricultural biogas plants in Poland, depending on the substrate use model. Energy Rep. **9**, 196–203 (2023). https://doi.org/10.1016/j.egyr.2023.05.175

W. Czekała, A. Jeżowska, D. Chełkowski, The use of biochar for the production of organic fertilizers. J. Ecol. Eng. **20**(1), 1–8 (2019). https://doi.org/10.12911/22998993/93869

W. Czekała, Agricultural biogas plants as a chance for the development of the agri-food sector. J. Ecol. Eng. **19**(2), 179–183 (2018). https://doi.org/10.12911/22998993/83563

W. Czekała, Biogas as a sustainable and renewable energy source, in *Clean Fuels for Mobility. Energy, Environment, and Sustainability*, ed. by G. Di Blasio, A.K. Agarwal, G. Belgiorno, P.C. Shukla (Springer, Singapore, 2022b). https://doi.org/10.1007/978-981-16-8747-1_10

J. Fulton, H. Cooley, P.H. Gleick, Water footprint, in *The World's Water. The World's Water*, ed. by P.H. Gleick (Island Press, Washington, DC, 2014). https://doi.org/10.5822/978-1-61091-483-3_5

G. Garcia-Garcia, E. Woolley, S. Rahimifard, J. Colwill, R. White, L. Needham, A methodology for sustainable management of food waste. Waste Biomass Valor **8**, 2209–2227 (2017). https://doi.org/10.1007/s12649-016-9720-0

M.L.M.B.B. Gonçalves, G.J. Maximo, Circular economy in the food chain: production, processing and waste management. Circ. Econ. Sust. (2022). https://doi.org/10.1007/s43615-022-00243-0

J.B.S.O.A. Guerra, M. Hoffmann, R.T. Bianchet, P. Medeiros, A.P. Provin, R. Iunskovski, Sustainable development goals and ethics: building "the future we want." Environ. Dev. Sustain. **24**, 9407–9428 (2022). https://doi.org/10.1007/s10668-021-01831-0

S. Hashim, M. Waqas, R.P. Rudra, A.A. Khan, A.A. Mirani, T. Sultan, F. Ehsan, M. Abid, M. Saifullah, On-farm composting of agricultural waste materials for sustainable agriculture in Pakistan. Scientifica 5831832 (2022). https://doi.org/10.1155/2022/5831832

R. H-Hargitai, V. Somogyi, Impact of water as raw material on material circularity—A case study from the Hungarian food sector. Heliyon **9**(7), e17587 (2023). https://doi.org/10.1016/j.heliyon.2023.e17587

A.Y. Hoekstra, A.K. Chapagain, Water footprints of nations: water use by people as a function of their consumption pattern, in *Integrated Assessment of Water Resources and Global Change*, ed. by E. Craswell, M. Bonnell, D. Bossio, S. Demuth, N. Van De Giesen (Springer, Dordrecht, 2006). https://doi.org/10.1007/978-1-4020-5591-1_3

S. Jørgensen, L.J.T. Pedersen, The circular rather than the linear economy, in *RESTART Sustainable Business Model Innovation. Palgrave Studies in Sustainable Business in Association with Future Earth* (Palgrave Macmillan, Cham, 2018). https://doi.org/10.1007/978-3-319-91971-3_8

Z. Kabir, M. Kabir, Solid waste management in developing countries: towards a circular economy, in *Handbook of Solid Waste Management*, ed. by C. Baskar, S. Ramakrishna, S. Baskar, R.

Sharma, A. Chinnappan, R. Sehrawat (Springer, Singapore, 2022). https://doi.org/10.1007/978-981-16-4230-2_1

M. Kameni Nematchoua, J.A. Orosa, Low carbon emissions and energy consumption: a targeted approach based on the life cycle assessment of a district. Waste **1**, 588–611 (2023). https://doi.org/10.3390/waste1030035

M. Kaur, A.K. Singh, A. Singh, Bioconversion of food industry waste to value added products: current technological trends and prospects. Food Biosci. **55**, 102935 (2023). https://doi.org/10.1016/j.fbio.2023.102935

N.J. Kennard, Food waste management, in *Zero Hunger. Encyclopedia of the UN Sustainable Development Goals*, ed. by W. Leal Filho, A.M. Azul, L. Brandli, P.G. Özuyar, T. Wall (Springer, Cham, 2020). https://doi.org/10.1007/978-3-319-95675-6_86

R. Khedkar, K. Singh, Food industry waste: a panacea or pollution hazard? in *Paradigms in Pollution Prevention. SpringerBriefs in Environmental Science*, ed. by T. Jindal (Springer, Cham, 2018). https://doi.org/10.1007/978-3-319-58415-7_3

D. Kim, C. Phae, Analysis of the effect of the use of food waste disposers on wastewater treatment plant and greenhouse gas emission characteristics. Water **15**, 940 (2023). https://doi.org/10.3390/w15050940

A. Lahiri, S. Daniel, R. Kanthapazham, R. Vanaraj, A. Thambidurai, L.S. Peter, A critical review on food waste management for the production of materials and biofuels. J. Hazard. Mater. **10**, 100266 (2023). https://doi.org/10.1016/j.hazadv.2023.100266

T. Mahmood, N. Hussain, A. Shahbaz, S.I. Mulla, H.M.N. Iqbal, M. Bilal, Sustainable production of biofuels from the algae-derived biomass. Bioprocess Biosyst. Eng. **46**, 1077–1097 (2023). https://doi.org/10.1007/s00449-022-02796-8

M.M. Mekonnen, W. Gerbens-Leenes, The water footprint of global food production. Water **12**, 2696 (2020). https://doi.org/10.3390/w12102696

S. Mokrane, E. Buonocore, R. Capone, P.P. Franzese, Exploring the global scientific literature on food waste and loss. Sustainability **15**, 4757 (2023). https://doi.org/10.3390/su15064757

B.G. Mwanza, C. Mbohwa, Sustainable solid waste management: a critical review, in *Sustainable Technologies and Drivers for Managing Plastic Solid Waste in Developing Economie*s. SpringerBriefs in Applied Sciences and Technology (Springer, Cham, 2022). https://doi.org/10.1007/978-3-030-88644-8_1

B. Obuobi, Y. Zhang, G. Adu-Gyamfi, E. Nketiah, Households' food waste behavior prediction from a moral perspective: a case of China. Environ. Dev. Sustain. (2023). https://doi.org/10.1007/s10668-023-03136-w

J. O'Connor, B.S. Mickan, K.H.M. Siddique, J. Rinklebe, M.B. Kirkham, N.S. Bolan, Physical, chemical, and microbial contaminants in food waste management for soil application: a review. Environ. Pollut. **300**, 118860 (2022). https://doi.org/10.1016/j.envpol.2022.118860

O.O. Olatunji, P.A. Adedeji, N. Madushele, Z.Z. Rasmeni, N.J. van Rensburg, Evolutionary optimization of biogas production from food, fruit, and vegetable (FFV) waste. Biomass Conv. Bioref. (2023). https://doi.org/10.1007/s13399-023-04506-0

M. Piao, A. Li, H. Du, Y. Sun, H. Du, H. Teng, A review of additives use in straw composting. Environ. Sci. Pollut. Res. **30**, 57253–57270 (2023). https://doi.org/10.1007/s11356-023-26245-5

A.K. Rashwan, H. Bai, A.I. Osman, K.M. Eltohamy, Z. Chen, H.A. Younis, A. Al-Fatesh, D.W. Rooney, P.-S. Yap, Recycling food and agriculture by-products to mitigate climate change: a review. Environ. Chem. Lett. (2023). https://doi.org/10.1007/s10311-023-01639-6

G. Shen, Z. Li, T. Hong, X. Ru, K. Wang, Y. Gu, J. Han, Y. Guo, The status of the global food waste mitigation policies: experience and inspiration for China. Environ. Dev. Sustain. (2023). https://doi.org/10.1007/s10668-023-03132-0

A. Siddiqua, J.N. Hahladakis, W.A.K.A. Al-Attiya, An overview of the environmental pollution and health effects associated with waste landfilling and open dumping. Environ. Sci. Pollut. Res. **29**, 58514–58536 (2022). https://doi.org/10.1007/s11356-022-21578-z

Z. Siddiqui, D. Hagare, M.-H. Liu, O. Panatta, T. Hussain, S. Memon, A. Noorani, Z.-H. Chen, A food waste-derived organic liquid fertiliser for sustainable hydroponic cultivation of lettuce. Cucumber Cherry Tomato. Foods **12**, 719 (2023). https://doi.org/10.3390/foods12040719

L. Susskind, J. Chun, A. Gant, C. Hodgkins, J. Cohen, S. Lohmar, Sources of opposition to renewable energy projects in the United States. Energy Policy **165**, 112922 (2022). https://doi.org/10.1016/j.enpol.2022.112922

E. Tamburini, M. Gaglio, G. Castaldelli, E.A. Fano, Biogas from agri-food and agricultural waste can appreciate agro-ecosystem services: the case study of Emilia Romagna region. Sustainability **12**, 8392 (2020). https://doi.org/10.3390/su12208392

C. Torres-León, N. Ramírez-Guzman, L. Londoño-Hernandez, G.A. Martinez-Medina, R. Díaz-Herrera, V. Navarro-Macias, O.B. Alvarez-Pérez B. Picazo, M. Villarreal-Vázquez, J. Ascacio-Valdes, C.N. Aguilar, Food waste and byproducts: an opportunity to minimize malnutrition and hunger in developing countries. Front. Sustain. Food Syst. **2**, 52 (2018). https://doi.org/10.3389/fsufs.2018.00052

M.F. Tröster, Assessing the value of organic fertilizers from the perspective of EU farmers. Agriculture **13**, 1057 (2023). https://doi.org/10.3390/agriculture13051057

F. Xu, Y. Li, X. Ge, L. Yang, Y. Li, Anaerobic digestion of food waste—Challenges and opportunities. Bioresour. Technol. **247**, 1047–1058 (2018). https://doi.org/10.1016/j.biortech.2017.09.020

T. Yigitcanlar, M. Kamruzzaman, M. Foth, J. Sabatini-Marques, E. da Costa, G. Ioppolo, Can cities become smart without being sustainable? A systematic review of the literature. Sustain. Cities Soc. **45**, 348–365 (2019). https://doi.org/10.1016/j.scs.2018.11.033

Printed in the United States
by Baker & Taylor Publisher Services